CW00794328

Britain and the Arctic

"The global gaze is hardening on the Arctic. Climate change is transforming the region, sovereignty issues and resource development are provoking geopolitical debate on its future, and indigenous peoples are asserting their rights and demanding greater involvement in decisions that affect their lives and lands. Meanwhile, a number of non-Arctic states are shaping their own approaches to the high latitudes. Depledge traces Britain's efforts to establish its own role in the Arctic; history, science, trade, conservation and national security are entangled with narratives about claims for a powerful presence in northern affairs. Erudite, incisive and original, this book is a vital contribution to scholarship on the contemporary Arctic and to our understanding of how the region is being redefined and contested by an array of interests."

—Mark Nuttall, *Professor and Henry Marshall Tory Chair, University of Alberta*

"This thought-provoking book considers the power-geometries of the Arctic Council and the exclusionary politics through which Arctic states attempt to deprive non-Arctic states of a say in regional affairs. Britain is the illustrative case, but the analysis is universally valid and applicable to other states as well. This makes the book a most important contribution also in the looming debate on how to improve on the legitimacy of Arctic decision-making in the future. For scholars and policy makers—Arctic and non-Arctic—this book is a MUST read."

—Willy Østreng, *President of the Norwegian Scientific Academy for Polar Research, 2012–2017*

"The idea that Britain is a 'forgotten' Arctic state surprises at first, but Depledge's explanation—close connections between the English and the Norsemen in the Viking Age, English fleets searching for northern sea routes to Asia, mass-scale whaling off Svalbard, Canadian Arctic territories under British rule, radioactive waste from Sellafield in the Barents Sea, the presence of British submarines in the Arctic Ocean, British funding of Arctic research—makes clear Britain's past and present proximity to the region."

—Lassi Heininen, *Professor and Leader of the Thematic Network on Geopolitics and Security and Co-founder of the GlobalArctic Project, University of Lapland*

Duncan Depledge

Britain and the Arctic

palgrave
macmillan

Duncan Depledge
Fleet, UK

ISBN 978-3-319-69292-0 ISBN 978-3-319-69293-7 (eBook)
https://doi.org/10.1007/978-3-319-69293-7

Library of Congress Control Number: 2017959056

© The Editor(s) (if applicable) and The Author(s) 2018
This work is subject to copyright. All rights are solely and exclusively licensed by the Publisher, whether the whole or part of the material is concerned, specifically the rights of translation, reprinting, reuse of illustrations, recitation, broadcasting, reproduction on microfilms or in any other physical way, and transmission or information storage and retrieval, electronic adaptation, computer software, or by similar or dissimilar methodology now known or hereafter developed.
The use of general descriptive names, registered names, trademarks, service marks, etc. in this publication does not imply, even in the absence of a specific statement, that such names are exempt from the relevant protective laws and regulations and therefore free for general use.
The publisher, the authors and the editors are safe to assume that the advice and information in this book are believed to be true and accurate at the date of publication. Neither the publisher nor the authors or the editors give a warranty, express or implied, with respect to the material contained herein or for any errors or omissions that may have been made. The publisher remains neutral with regard to jurisdictional claims in published maps and institutional affiliations.

Cover illustration: © Melisa Hasan

Printed on acid-free paper

This Palgrave Macmillan imprint is published by Springer Nature
The registered company is Springer International Publishing AG
The registered company address is: Gewerbestrasse 11, 6330 Cham, Switzerland

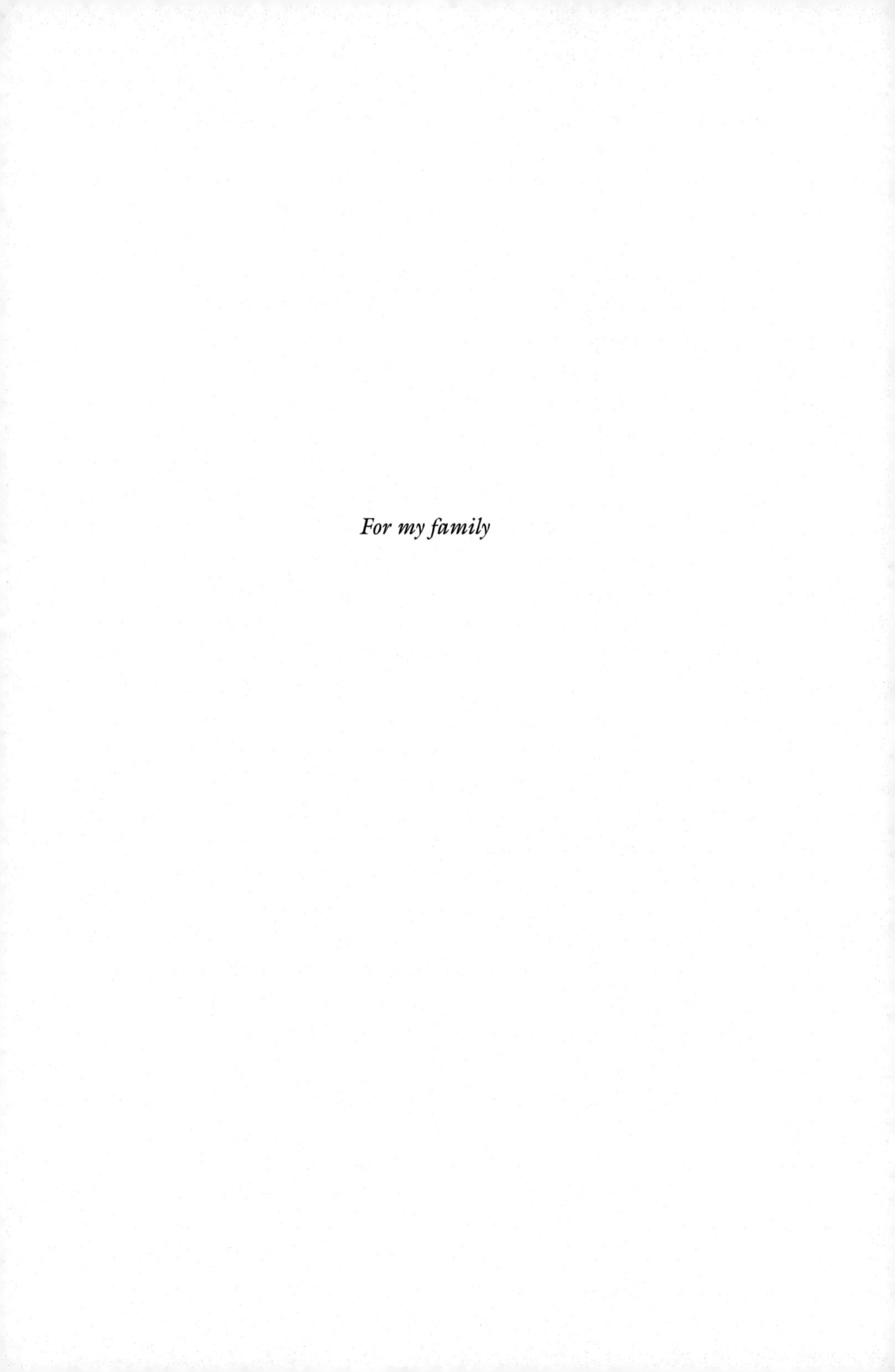

For my family

Preface

This book grew out of the doctoral project—funded by the Royal United Services Institute and the Economic and Social Research Council—that I completed at Royal Holloway, University of London, between 2010 and 2014, entitled *Being Near-Arctic: A Critical Geopolitics of Contemporary British Policy Towards the Far North*. After successfully passing my viva, my two examiners encouraged me to turn my study into a book, not least because it would be the first substantial title to address Britain's geopolitical interests in the Arctic since the end of the Cold War. It would also be timely as British interest in the evolving geopolitics of the Arctic has undoubtedly grown over the past decade.

In the three years that passed since I submitted my dissertation, writing this book further provided me with an opportunity to reflect on, rethink, and update my original study. When I sat down for my viva in 2014, the House of Lords Select Committee was only a few months into its wide-ranging inquiry into the opportunities and risks arising in the Arctic, and their implications for Britain. Since then we have seen a new crisis in West–Russia relations, a dramatic fall in oil prices, and record high temperatures in the Arctic, all of which have had a bearing on how we should think about British interests in the region.

I have also been fortunate enough to become more deeply acquainted with Britain's various Arctic stakeholders in Government, Parliament, the military, the private sector, academia, and civil society, both in my capacity as Director of the All-Party Parliamentary Group for Polar Regions Secretariat (since 2015), and while serving as special adviser to the House of Commons Defence Committee during the recent sub-Committee inquiry into Defence in the Arctic (since 2017). These acquaintances have without doubt helped me to reflect more deeply on how British interests in the Arctic have evolved.

The research methods and the theoretical underpinnings of this book are the same as those described in my dissertation, which is openly accessible from Royal Holloway. Over the past seven years I have regularly found myself in close proximity to those actually making and shaping British policy towards the Arctic, for instance, as a participant in the 2010 Canada–UK Colloquium, which focussed that year on the Arctic, as an observer of British military forces during Exercise COLD RESPONSE in 2012, as a participant in the Ministry of Defence's study group on the Polar Regions in 2013, as a member of the British delegation to the Arctic Circle Assembly in Reykjavik in 2014, as a participant in the Wilton Park Conference on the future of the Arctic in 2016, as well as through my work with the All-Party Parliamentary Group for Polar Regions, and the House of Commons Defence Committee. Much of my approach was opportunistic, snatching where I could glimpses of British Arctic policy debates at meetings, workshops, and conferences, and supplementing those with more than 50 interviews with Parliamentarians, current and former civil servants and Government scientists, representatives of civil society and the private sector, journalists, and other academics. I have not been able to reveal everything I heard and saw, but I believe that was a price worth paying for the insights that I accumulated.

Lastly, the project is unapologetically policy-focused and Whitehall-centric. There are still questions to be asked about how the wider public thinks about the Arctic today, in contrast to our Victorian predecessors who continue to receive substantially more attention, but that could not be included within the scope of either my doctoral project or this book. There are also outstanding questions about the extent to which Britain's interests in the Arctic diverge depending on where you are in the country. For instance, the Scottish Government has complained that its particular interests in the Arctic continue to be ignored, while other parts of Britain, including cities such as Hull, might also feel that they have a distinct stake in Arctic affairs. Far from offering the final word, I hope this book will mark the beginning of a more diverse debate about what the Arctic means to Britain, and how Britain should engage with the Arctic in all its diversity, in the twenty-first century.

Acknowledgements

This book probably would not have happened if not for the encouragement I received from my doctoral examiners Professor Clive Archer and Dr Richard Powell. I was particularly saddened to learn of Professor Archer's passing in 2016, before I could bring this book to fruition.

I am indebted to those people who have taken the time to talk to me about the Arctic over the past seven years, allowing me to benefit from all their accumulated knowledge and experience, although none bear responsibility for my subsequent analysis of what I learned. There are simply too many to list them all by name.

I am immensely grateful to a host of other individuals (including my entire family) who have at some stage or other supported me with the writing of this book, whether by offering their precious time to respond to my persistent questioning and offer comments on specific chapters or sections of the book, or simply through their gentle support and encouragement. Special mention must go to Dr Andrew Foxall, Dr Dougal Goodman, Michael Kingston, Professor Tim Benton, Professor Michel Kaiser, Professor Peter Wadhams, Professor Ray Leakey, Professor Terry Callaghan, Dr Kathrin Keil, Henry Burgess, Nick Cox, James Rogers, Matt Skuse, and James Gray, MP.

I also thank Joanna O'Neill and the team at Palgrave for their patience and assistance, Jenny Kynaston in the Department of Geography at Royal Holloway for helping me to secure the map that is included in this book, and Matthew Gale for putting together the index.

Finally, I cannot thank Professor Klaus Dodds enough for both his guidance and his friendship since our chance meeting in 2009. He has gone above and beyond his responsibilities as my doctoral supervisor—which ended three years ago—with the support he has given me throughout this project.

The map of the Arctic included in this book was modified from MountainHighMapsPlus® Copyright ©2013 Digital Wisdom Publishing Ltd.

Contents

List of Abbreviations

ABA	Arctic Biodiversity Assessment
ACIA	Arctic Climate Impact Assessment
AEPS	Arctic Environmental Protection Strategy
AMEC	Arctic Military Environmental Cooperation
AMSA	Arctic Marine Shipping Assessment
ARR	Arctic Resilience Report
ASW	Anti-submarine warfare
BAS	British Antarctic Survey
BBC	British Broadcasting Service
CLCS	Commission on the Limits of the Continental Shelf
DECC	Department of Energy and Climate Change
EEZ	Exclusive Economic Zone
EU	European Union
GIUK	Greenland–Iceland–UK Gap
IASC	International Arctic Science Committee
IGY	International Geophysical Year
IMO	International Maritime Organisation
IPCC	Intergovernmental Panel on Climate Change
IPY	International Polar Year
LNG	Liquefied Natural Gas
MOU	Memorandum of Understanding
MP	Member of Parliament
NARF	National Arctic Research Forum
NATO	North Atlantic Treaty Organization
NERC	Natural Environment Research Council

NGO	Non-governmental Organisation
PSC	Polar Sciences Committee
RGS	Royal Geographical Society
SAMS	Scottish Association for Marine Sciences
SCAR	Scientific Committee on Antarctic Research
SDSR	Strategic Defence and Security Review
SPRI	Scott Polar Research Institute
SLBM	Submarine-launched ballistic missile
UKTI	UK Trade & Industry
UN	United Nations
UNCLOS	United Nations Convention on the Law of the Sea
UNFCCC	United Nations Framework Convention on Climate Change
USGS	United States Geological Survey
WWF	World Wide Fund for Nature

CHAPTER 1

Introduction: Britain and the Arctic

Abstract Britain's interest in the Arctic is at its highest level since the end of the Cold War. As the Arctic Ocean undergoes a profound state change from being permanently ice-covered to seasonally ice-free, British policy-makers, businesses, scientists, and civil society have all entered the global scramble to redefine why the Arctic matters. This chapter introduces Britain's stake in the Arctic and the challenges it faces in making its contemporary interests in the region heard.

Keywords Britain and the Arctic • Climate change • Global Arctic • Circumpolarisation • Contested Arctic • Proximity

In April 2006, the leader of the British Conservative Party, David Cameron, visited the Norwegian Arctic as part of a trip supported by the World Wide Fund for Nature (WWF). During the trip, several iconic photographs emerged of Cameron riding a husky-powered sled as he visited a remote Norwegian glacier and saw for himself the effects of climate change on the Arctic. As Cameron urged voters to 'vote blue to go green' in local elections back in Britain, he wanted to show the British public that he understood their concerns about climate change. How better to do so than by choosing the Arctic, which is warming twice as fast as anywhere else on Earth, for his first major trip after becoming party leader. However, Cameron's record on climate change and environmental issues during his time as Prime Minister (2010–2016) left many doubting whether his visit

© The Author(s) 2018

D. Depledge, *Britain and the Arctic*,
https://doi.org/10.1007/978-3-319-69293-7_1

to the Arctic was anything more than a publicity stunt. Instead, Cameron was criticised for using the Arctic, and the vulnerability that it represents to environmentalists, to get elected, only to later renege on his commitment to lead the 'greenest government ever'. For Cameron, it seemed, the Arctic was merely another place for performing domestic politics (Fig. 1.1).

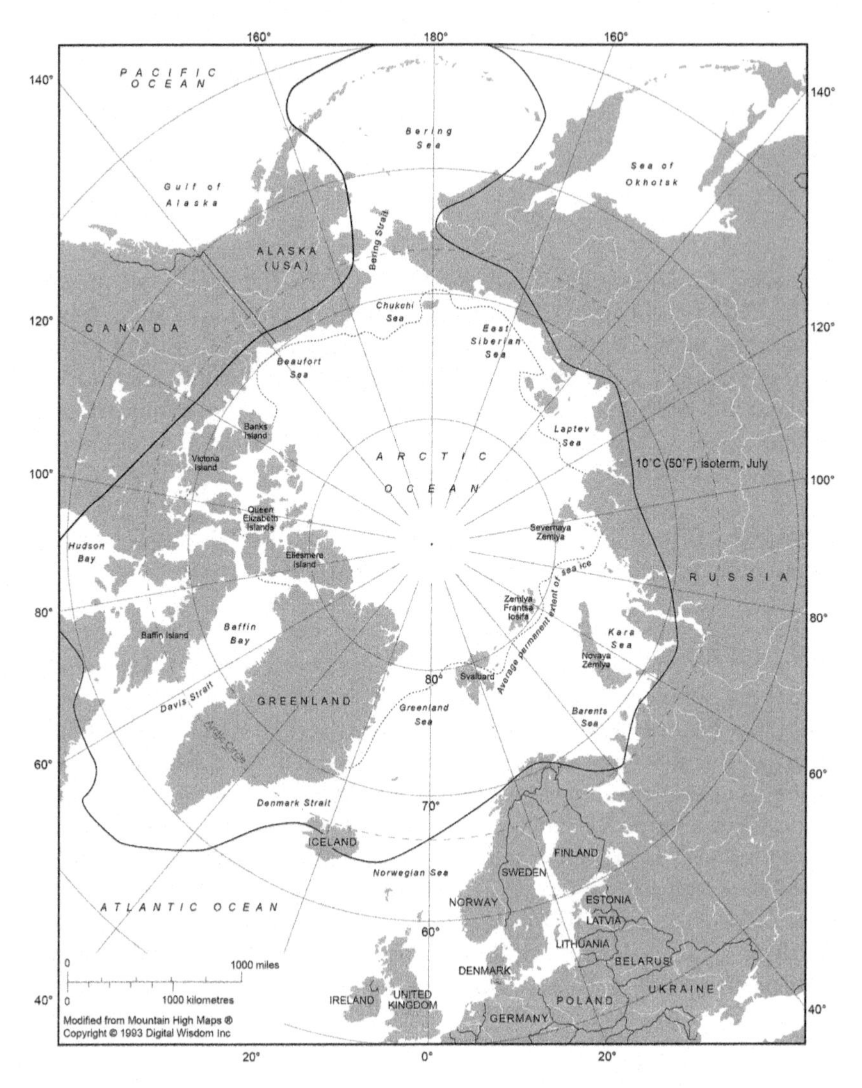

Fig. 1.1 The Arctic

* * *

In 2015, the House of Lords Select Committee on the Arctic (hereafter, Arctic Committee) called on the British Government to appoint an Ambassador to the Arctic (Arctic Committee 2015). Diplomatically speaking, that would be considered an exceptional act: Britain appoints Ambassadors to countries, not regions. Yet as other countries such as Finland, France, Japan, and Singapore went about appointing their own 'Arctic' or 'Polar' Ambassadors, there seemed to be something about the Arctic that attracted exceptional acts and exceptional interest (Dittmer et al. 2011).

The Arctic is still, after all, one of the world's most extreme, least understood, and inaccessible environments. In much of the Arctic, it is enormously expensive to deploy the nationalistic, commercial, scientific, and militaristic means of geopolitical intervention that have tended to define interstate competition elsewhere in the world. For instance, during the Cold War, only the United States, the Soviet Union, and Britain could afford to engage in cat-and-mouse submarine warfare under the ice, and only then because control of the Arctic promised a safe haven from which to launch a nuclear attack, while denying opponents the same. When the Cold War ended, all three countries substantially reduced their operations there. In the years that followed, international scientific cooperation, and institution-building to support it, took centre stage in a way largely unseen in other parts of the world. Except, perhaps, in Antarctica, where similarly, no single nation had the political will or economic resources to act alone.

Political events may have been the primary driver of geopolitical events in the Arctic in the twentieth century, but awareness is now growing of the profound environmental changes that have also been under way in the region since 1979, when satellite monitoring of Arctic sea ice started. The summer sea ice minimum has fallen from around 6 to 7 million square kilometres in the 1980s and 1990s, to between 3 and 5 million square kilometres since 2007, an average decline of 13.4% per decade. Low summer minima have subsequently become the norm, with the ten lowest on record all occurring since 2005. More widely, there is no longer any part of the Arctic where the Arctic sea ice coverage is greater than it was during the 1980s and 1990s (NASA 2017).

As the Arctic Ocean transitions from being permanently ice-covered to seasonally ice-free, the possibilities of human activity there are being restructured (Berkman and Young 2009). The extent of Arctic environmental change was demonstrated in 2016, when the Polar Ocean Challenge team led by the British explorer Sir David Hempleman-Adams sailed an aluminium-hulled yacht through the Northeast and Northwest Passages in a single summer, encountering only modest ice conditions along the way. In that same summer, the wrecks of two British ships—HMS *Terror* and *The Thames*—were discovered, nearly two centuries after they had foundered in thick sea ice in those very same passages. As this book goes into production, the British explorer Pen Hadow, is leading another team in an attempt to sail two yachts between cracks in the ice to the geographic North Pole.

Others have responded to Arctic environmental change by variously imagining an ice-free Arctic as a front line for climate change, a commercial frontier, a strategic theatre, an increasingly populated homeland, or a protected nature reserve (Dodds 2010; Steinberg et al. 2015). Interest in the Arctic has also become more widespread. While explorers, merchants, scientists, and states have long been interested in the Arctic, they mostly stemmed from Europe, and later North America. Today, several Arctic scholars write of a 'Global Arctic', to analyse and examine the increasing diversity of contemporary interest, including from countries such as Brazil, China, Japan, Indonesia, Malaysia, Singapore, South Korea, and Vietnam (Heininen and Finger Forthcoming). Part of their interest is in how changes in the Arctic environment will affect sea levels and weather patterns in temperate zones. Simultaneously, however, they are also looking whether it will become easier for actors from beyond the region to traverse the Arctic and harvest the region for resources. Both perspectives were on display at the World Economic Forum meeting in Davos in 2017 when leading climate scientists and senior political figures set out the global economic risks of Arctic change.

Meanwhile, the international news media has tended to exaggerate the extent to which Arctic environmental change is driving interstate competition by claiming that a new 'Great Game' is under way, as if the Arctic is being subjected to the same imperial machinations that Africa was in the nineteenth century. However, the struggle for influence in the Arctic is different. Contemporary contests are better described as a scramble[1] between different constellations of state and non-state actors to define what kind of place the Arctic is becoming as it warms, loses ice and permafrost, greens, and unwittingly becomes host to alien species and geopolitical intrigue.

When, in 1921, Frank Debenham, who went on to become the first director of the Scott Polar Research Institute (SPRI) in Cambridge, wrote about 'The Future of Polar Exploration' he included a map in which much of the Arctic was simply labelled 'unexplored' (Debenham 1921). Less than 100 years ago, much of the Arctic was still unknown, at least to Western science, business, and industry. Today, despite significant advances in knowledge and understanding, as climate change transforms the region, the Arctic is once again becoming a blank space in our mental maps of the world, a space which is dominated by uncertainty and lack of knowledge about the risks and opportunities that might be found there. As the political geographer Richard Powell (2008: 827) observed during the media hysteria that surrounded the Arctic in 2008:

> The Arctic Ocean has again become a zone of contestation. In this contemporary clash of scientific knowledges, legal regimes and offshore technologies, the uncertain spatialities of the Circumpolar Regions are being reconfigured.

There are, then, several possibilities in play that different constellations of actors, involving Arctic as well as non-Arctic, state as well as non-state, will have a say in defining whether the Arctic comes to be imagined as a 'New North' of economic enterprise tied into global commerce, a homeland for indigenous peoples, an environmental sanctuary, a strategic theatre, or the harbinger of global climate terror—and consequently, whether the Arctic is seen as a place to be occupied, harvested for resources, militarised, or protected from human activity.

However, in this scramble to redefine the Arctic, the kinds of nationalistic, commercial, scientific, and military enterprises witnessed in the Arctic for several centuries have largely given way to more cooperative means involving international scientific programmes, institution-building and international law, and joint commercial ventures. Today, a country's influence in Arctic affairs is arguably far more likely to be defined by its ability to shape the form and direction of those international programmes, shared institutions, and commercial ventures than it is by nationalistic means. The Arctic Committee's call for Britain to appoint an Ambassador for the Arctic, who could represent a wide range of British interests to such programmes, institutions, and ventures spoke directly to this changed reality by proposing that an exceptional type of diplomatic intervention, to serve as a bridge between Britain and the Arctic, was needed to bolster Britain's influence in the region.

The New (and Contested) Arctic

British interest in the Arctic today is based around the need to comprehend how the Arctic is changing, and the related desire to put any new knowledge to work in ways which are productive in terms of science, trade, conservation, and national security. That interest is widespread, encompassing stakeholders from civil service, industry, national research centres, academia, and civil society. However, both the Arctic's diversity and attempts to fathom it have been complicated by interactions between climate change and pre-existing environmental differences, as well as the uneven spread of pollution and human development across the region. As the Arctic Climate Impact Assessment (ACIA) showed, climate change is impacting everything from sea ice, to ecosystems, people, and, ultimately, the prospects of life itself across the region (ACIA 2004). Old knowledges are melting away, and new knowledges are needed in their place.

Simultaneously, the Arctic has come to be populated by a much wider array of actors interested in accessing, inhabiting, studying, testing, harvesting, and mining it in support of a variety of interests. While this has, to some extent, always been the case, perhaps the crucial difference is that many more actors are now being heard. The power geometries[2] have shifted to the point where the Old European powers, and, more recently, the Arctic states, are no longer able to monopolise the projection of Arctic imaginaries, at the expense of all others. Indigenous peoples, non-governmental organisations (NGOs), new industries (such as the renewable energy sector), and non-Arctic states such as China, Singapore, and Japan are all offering their own imaginaries of what the Arctic should be.

Consequently, the first challenge facing Britain is to understand Arctic environmental change—a subject matter that is being addressed by an array of British-based physical, environmental, and social scientists funded by British research councils and the European Union (EU), and further supported through bilateral partnerships with countries such as Canada and Norway. However, Britain also has to grasp the contests which are playing out between a wide variety of actors to define the Arctic's future, and figure out how best to enrol others in ways that support its own interests in the region. That there have been five separate parliamentary inquiries on Britain's role and interests in Arctic affairs in the past five years alone suggests that this is no easy task. Each inquiry has taken a different point of departure, demonstrating that there is little consensus yet in

Westminster and Whitehall about what Britain's interests and role in the Arctic should be.

At the same time though, the literary/cultural postcolonial scholar Graham Huggan (2016) recently offered a reminder that there are colonial anxieties in play every time Britain utters anything about, or acts in, the Arctic. Those anxieties are felt by the indigenous peoples of the Arctic as well as the Arctic states themselves, which, as Chap. 3 shows, have spent much of the past century nation-building in the Arctic. What the feminist art historian Lisa Bloom and colleagues have noted in general certainly seems apt for Britain:

> Some of the same discursive strategies we are seeing now, particularly the way the Arctic is being re-imagined by drilling proponents of the oil and gas industry as a conveniently 'empty frozen wasteland of snow and ice' replay earlier imperial narratives of Arctic and Antarctic exploration in which those territories were imagined as 'white' or 'blank' spaces to be filled in by the very Europeans who designated them so. (Bloom et al. 2008: 1)

Moreover, Huggan (2016) has argued that, as such discourses are mapped onto the Arctic by political scientists and other designated Arctic experts (typically Western, but increasingly also Asian), there is a risk that the Arctic will be reinscribed as an object of Western/Asian knowledge, which in turn could be regarded as a form of epistemic violence[3] against postcolonial imaginaries and practices. For example, in 2009, the EU was accused by the then-Canadian Foreign Minister Lawrence Cannon of being insensitive and ignorant of the needs and interests of indigenous peoples, particularly in the Canadian Arctic and Greenland, when it pushed ahead with a decision to ban seal products from its markets (CBC News 2009, see also Wegge 2013). Meanwhile, others have been cautious about welcoming China into the Arctic out of fear that its sheer economic heft will be enough to buy it the influence it needs to play a significant role in remaking the Arctic in accordance with its own interests (Willis and Depledge 2015).

Approach

In recent attempts to understand how the Arctic is changing (e.g. by investing more than £30 million in national Arctic science programmes since 2009), and to put that knowledge to work in ways that create new diplomatic and commercial opportunities in the Arctic, the British Government risks pushing its own form of neocolonial imaginary in the Arctic. That risk is further enhanced by enduring memories of Britain's past history as an imperial power that has sought to project its influence and draw resources from virtually all corners of the Earth. The need to be sensitive to these anxieties has been important for shaping the development of contemporary British Government policy towards the Arctic. This book is therefore as much about how Britain is recalibrating its relations with the Arctic in response to both climate change and the postcolonial sensitivities that it must navigate as it is about Britain's contemporary interests in the Arctic.

Four themes are woven through the subsequent chapters. The first theme is that Britain has had a role in defining the Arctic for centuries, beginning with the attempts of Elizabethan explorers, such as Martin Frobisher, to find maritime passages through the ice from the Atlantic to the Pacific, which would render the Arctic traversable like any other ocean. The actions and ideas of the Crown, Parliament, explorers, scientists, whalers, sealers, fur trappers, and merchants also shaped ideas about Britain's relative proximity to the region, irrespective of the actual physical distances involved. That in turn poses a provocative question about why it is only recently that Britain has started defining itself as the Arctic's 'nearest neighbour'.

The second theme is that despite the increasing 'circumpolarisation' of Arctic affairs in the twentieth century, whereby the so-called Arctic states (Canada, Denmark, Finland, Iceland, Norway, Russia, Sweden, and the United States) have pushed non-Arctic states such as Britain towards the periphery of debates about the region's future, the region continues to attract attention from British Government officials, military personnel, industry, academia, and civil society, primarily as a consequence of the dramatic environmental changes under way in the region, and a desire to shape the region's future. As such, much of what the British Government is doing when it claims an interest in the Arctic is challenging the existing trajectories of circumpolarisation by intervening with alternatives that seek to remake and reinterpret Arctic geographies in ways that bring Britain and the Arctic closer together.

The third theme is that as existing knowledge of the Arctic melts away, attempts by Britain to comprehend what is happening in the region, and to use this understanding for productive ends, are invariably speculative, and informed by anticipatory logics that seek to position Britain in such a way that it can be responsive to a variety of Arctic futures (whether those futures involve greater demand for science, industry, environmental protection, or military operations). For example, the British Government's investment in new Arctic science over the past decade is a sure sign that there remains plenty of interest in better understanding the dynamics of sea ice loss, ocean acidification, and other emerging issues.

The fourth theme is that contemporary British engagement with the Arctic, as a consequence of Britain's own imperial past in the Arctic (which continues to be felt, for instance, in the names that populate maps of the Arctic, in the discoveries of nineteenth-century shipwrecks, and in the lionising and memorialising of explorers such as Sir Martin Frobisher and Sir John Franklin), is also shaped by its need to demonstrate sensitivity to ongoing debates about postcolonialism and neocolonialism. At times, as this book shows, this has at least in part led successive governments since the end of the Cold War to be overcautious about adopting an openly proactive policy towards the Arctic. More recently, there was much hand-wringing among Foreign and Commonwealth Office (hereafter, Foreign Office) officials over whether the British Government should produce a white paper on the Arctic as evidenced by the responses they gave under questioning by the Environmental Audit Committee (2012–2013) and the Arctic Committee (2014–2015). At other times, it has served as a reminder of the constraints on Britain's ability to influence a region now largely under the sovereign jurisdiction of other countries and peoples.

Given the emphasis on these themes, academic audiences should view the book as situated within the emerging, self-styled field of Critical Polar Geopolitics.[4] Such an approach encourages us to think about the ways in which geographies of the Arctic (and Antarctica) remain unsettled, even scrambled (Powell and Dodds 2014). In other words, attempts to define what these spaces are, how they are to be rendered productive, and who is to be involved in these processes are always ongoing, meaning the future of the Arctic (including who or what constitutes the Arctic, broadly speaking) remains open. The international legal scholar Timo Koivurova's description of the Arctic as a 'region-in-change' remains particularly powerful in this regard because it raises the prospect that 'traditional' Arctic actors such as Britain need to stay invested in the various processes that are redefining the Arctic and its governance structures, or else their own status as 'Arctic' or 'near-Arctic' actors may be called into question (Koivurova 2010: 153).

Whether this investment is being retained and reinforced by the British Government is central to this book.

Such an approach stands in marked contrast to the work of previous British-based polar scholars, who have tended to treat the Arctic as a region to be surveyed, sorted, and catalogued, primarily in terms of its material characteristics, including physical geography, economic productivity, demography, forms of industry, and hard political boundaries (Armstrong et al. 1978; Archer and Scrivener 1986). An earlier intervention by the Indian political scientist Sanjay Chaturvedi (1996), who was at the time a resident scholar at SPRI, was one of the first to bring a critical geopolitical perspective to the study of the polar regions—but it contained little commentary on how these issues related specifically to Britain.

Broadly speaking, then, to the author's knowledge, this book is the first to investigate why the Arctic matters to Britain in the early twenty-first century and beyond.

Notes

1. Here, the term 'scramble' is informed by Dodds and Nuttall's (2016) work highlighting the ways in which the Arctic and Antarctica are defined by creativity and uncertainty.
2. The eminent geographer Doreen Massey (1994) used the term 'power geometry' to make the point that the relations that govern flows and movements (of goods, people, ideas, practices, etc.) between different social groups are rarely even. Some social groups are more in charge of these movements than others and as such are able to use their encounters with other social groups to increase their power and influence, while at the same time reducing that of others.
3. A term used to describe how violence can be inscribed against others through discourse.
4. On Critical Geopolitics in general see the seminal work by Gearóid Ó Tuathail (1996). For a shorter introduction see Klaus Dodds (2014).

References

ACIA. 2004. *Impacts of a Warming Arctic: Arctic Climate Impact Assessment.* Cambridge: Cambridge University Press.

Archer, Clive, and David Scrivener. 1986. Introduction. In *Northern Waters*, ed. Clive Archer and David Scrivener, 1–10. Totowa: Barnes & Noble.

Arctic Committee. 2015. *Responding to a Changing Arctic.* London: The Stationary Office Limited.

Armstrong, Terence, George Rogers, and Graham Rowley. 1978. *The Circumpolar North*. London: Methuen & Co Ltd.
Berkman, Paul, and Oran Young. 2009. Governance and Environmental Change in the Arctic Ocean. *Science* 324: 339–340.
Bloom, Lisa, Elena Glasberg, and Laura Kay. 2008. Introduction: New Poles, Old Imperialism? *Scholar and Feminist Online* 7: 1–6. Accessed October 27, 2016. http://sfonline.barnard.edu/ice/intro_01.htm
CBC News. 2009. Canada Against EU Entry to Arctic Council Because of Seal Trade Ban. *CBC News*, April 29. Accessed June 22, 2017. http://www.cbc.ca/news/canada/north/canada-against-eu-entry-to-arctic-council-because-of-seal-trade-ban-1.806188
Chaturvedi, Sanjay. 1996. *The Polar Regions: A Political Geography*. Chichester: John Wiley & Sons.
Debenham, Frank. 1921. The Future of Polar Exploration. *The Geographical Journal* 57: 182–200.
Dittmer, Jason, Sami Moisio, Alan Ingram, and Klaus Dodds. 2011. Have You Heard the One About the Disappearing Ice? Recasting Arctic Geopolitics. *Political Geography* 30: 202–214.
Dodds, Klaus. 2010. A Polar Mediterranean? Accessibility, Resources and Sovereignty in the Arctic Ocean. *Global Policy* 1: 303–310.
———. 2014. *Geopolitics: A Very Short Introduction*. 2nd ed. Oxford: Oxford University Press.
Dodds, Klaus, and Mark Nuttall. 2016. *The Scramble for the Poles*. Cambridge: Polity Press.
Heininen, Lassi, and Matthias Finger. Forthcoming. The 'Global Arctic' as a New Geopolitical Context and Method. *Journal of Borderlands Studies*.
Huggan, Graham. 2016. Introduction: Unscrambling the Arctic. In *Postcolonial Perspectives on the European High North*, ed. Graham Huggan and Lars Jensen, 1–29. Basingstoke: Palgrave Macmillan.
Koivurova, Timo. 2010. Limits and Possibilities of the Arctic Council in a Changing Scene of Arctic Governance. *Polar Record* 46: 146–156.
Massey, Doreen. 1994. *Space, Place and Gender*. Minneapolis, MN: University of Minnesota Press.
NASA. 2017. Arctic Sea Ice. *Earth Observatory*. Accessed July 3, 2017. https://earthobservatory.nasa.gov/Features/SeaIce/page3.php
Ó Tuathail, Gearóid. 1996. *Critical Geopolitics*. Minneapolis, MN: University of Minnesota Press.
Powell, Richard. 2008. Configuring an 'Arctic Commons'? *Political Geography* 27 (2008): 827–832.
Powell, Richard, and Klaus Dodds. 2014. Polar Geopolitics. In *Polar Geopolitics? Knowledges, Resources and Legal Regimes*, ed. Richard Powell and Klaus Dodds, 3–18. Cheltenham: Edward Elgar.

Steinberg, Philip, Jeremy Tasch, Hannes Gerhardt, Adam Keul, and Elizabeth A. Nyman. 2015. *Contesting the Arctic: Politics and Imaginaries in the Circumpolar North*. London: I.B. Tauris.

Wegge, Njord. 2013. Politics Between Science, Law and Sentiments: Explaining the European Union's Ban on Trade in Seal Products. *Environmental Politics* 22: 255–273.

Willis, Matthew, and Duncan Depledge. 2015. How We Learned to Stop Worrying About China's Arctic Ambitions: Understanding China's Admission to the Arctic Council, 2004–2013. In *Handbook of the Politics of the Arctic*, ed. Leif Christian Jensen and Geir Hønneland, 388–407. Cheltenham: Edward Elgar.

CHAPTER 2

Britain: The Forgotten Arctic State?

Abstract Today, it is widely accepted that there are eight Arctic states: Canada, Denmark, Finland, Iceland, Norway, Russia, Sweden, and the United States. Their identity is determined by their topographical geography, which encompasses territories north of the Arctic Circle. According to this topographical logic, Britain, too, was once an Arctic state, but it now refers to itself as the Arctic's 'nearest neighbour'. However, as this chapter argues, proximity is not solely about topography. It is also about topology—in other words, the extent to which Britain and the Arctic are folded together and connected by flows of bodies, knowledges, resources, and practices. In the absence of a contemporary Arctic topography, it is these topologies which give Britain a stake in the future of the region.

Keywords Britain and the Arctic • Proximity • Topology • Topography • Global Arctic • Arctic history

What is it to be an 'Arctic' state? Is Britain an Arctic state? Has it ever been? How is such a question answered? When so-called Arctic states Canada, Denmark, Finland, Iceland, Norway, Russia, Sweden, and the United States met in 1987 to start negotiations on the International Arctic Science Committee (IASC) they determined that a state could only be considered Arctic if it had sovereign jurisdiction over land north of the Arctic Circle (English 2013). The Rovaniemi Process to establish

© The Author(s) 2018

D. Depledge, *Britain and the Arctic*,
https://doi.org/10.1007/978-3-319-69293-7_2

a cooperative framework for addressing emerging environmental challenges in the region, launched by the Arctic states in 1989, reinforced that precedent. Ever since, the aforementioned countries have been referred to collectively as the Arctic states (or A8).

The A8 grouping is informed by a *topographical* view of geography. The world is imagined as something that is easily divided into discreet parts (Taylor 1994). Look at any standard map of the world and it is as if it has been put together like a jigsaw, with each state, region, sea, or ocean representing a separate piece. If we take the line of latitude that marks the Arctic Circle at approximately 66°N, we only need look at what land lies above it to determine which parts of the world are 'Arctic' and which are 'non-Arctic'. Accordingly, states with sovereign jurisdiction over 'Arctic' lands are generally regarded as having primacy over what happens there.

However, a *topological* view of geography produces a different outcome. Space is regarded as a continuum, rather than a set of discreet parts (Elden 2005). Inspired in part by the philosophical work of Gilles Deleuze and Felix Guattari (2004), thinking topologically about Arctic identity emphasises the ways in which seemingly discreet places fold into one another to create what some describe as a 'scrumpled' geography (Doel 1996). Put differently, if you scrunch up a map in your fist, seemingly distant places can be brought closer together. Similarly, when geography is regarded as being scrunched and crumpled together as a consequence of flows and connections created by the movement of people, materials, ideas, and practices between different countries and regions, and so on, places that may seem distant when viewed on a plane (i.e. the way geography is represented on most maps) are actually found to be much closer than previously imagined (Doel 1996).

Proximity—what we consider to be close or distant—therefore appears to have both a physical and an abstract dimension. Physically, the distance between the northern tip of the British Isles and the Arctic Circle is approximately 400 kilometres. However, whether we consider such a distance to be great or small depends more on abstract understandings of whether people, materials, ideas, and practices find it is easy and quick to traverse, and the regularity with which that happens. Our sense of proximity might also be affected by how exposed we feel to what happens 400 kilometres away. Throughout everyday life we find ourselves feeling closer to places, people, and materials which are physically further away. That is usually because familiarity and knowledge breeds a sense of proximity. Conversely, we tend to feel distant from those places, people, and materials that seem unfamiliar, or of which we lack knowledge and experience. Moreover, our sense of proximity can change over time, as our knowledge and experience evolves. In other words, the connections that make up a topological geography of connections and flows can shift, break, and form anew. That is why globalisation, increasing connectivity, and shorter time horizons are described as shrinking the world, while more recent claims that isolationism and protectionism are on the rise are based on arguments that in fact the opposite is occurring.

Both topographical and topological notions of geography must be discussed if we are to answer the question of whether Britain is a 'forgotten' Arctic state. Both ways of thinking are evident in contemporary formulations of Arctic policy by successive British governments between 2010 and 2017, which started to refer to Britain as the Arctic's 'nearest neighbour'.

The issue of forgetfulness is intended to be deliberately provocative as it forces Britain and others to engage with the fact that, for a period of nearly 400 years—a period longer than many of today's Arctic states have even existed as independent sovereign states—it was, by the contemporary definition, an Arctic state, as a consequence of its imperial holdings in North America, as well as its industrial history elsewhere in the Arctic in places such as Svalbard. It is a history which is conveniently 'forgotten' by the Arctic states when they attempt to root their own sense of Arctic sovereignty in a topography which is timeless and unchanging in the natural order of things. At the same time, the spread and intensity of Britain's connections to the Arctic (i.e. Britain's Arctic topology) has changed over the centuries, to the point where today, in spite of memorials, museums, and other traces of Britain's Arctic history which continue to haunt places as diverse as London, Bristol, Hull, and Dundee, many in Britain appear to have 'forgotten' Britain's Arctic history. Moreover, many also seem to continue 'forgetting' Britain's present proximity to what is happening in the region, despite the regularity of dire-sounding warnings in the tabloids that the country is about to be hit by blasts of 'Arctic weather', as well as more sobering reports about how climatic changes in the Arctic might affect Britain and its interests around the world.

Britain's ongoing forgetfulness was particularly evident when, in October 2016, the British Broadcasting Service (BBC) broadcast a three-part television programme called *Arctic Live* that interspersed short films about contemporary Arctic issues with live footage of polar bears. As viewers learned about demands for Arctic oil, Canadian military activity, and the need to protect precious wildlife, the producers seemed to forget that Britain's own military has operated near-continuously in the Arctic since the Second World War; that state-owned British Petroleum (now privately owned BP) was instrumental in developing Prudhoe Bay, the first major Arctic oil field in North America; that BP and Shell (both still British-based companies) continue to seek out new extractive opportunities in the region; and that British branches of Greenpeace and the World Wide Fund for Nature (WWF) have been central to mobilising global civil society support for protecting the Arctic environment.

The use of live footage of polar bears to entice British audiences into an encounter with the Arctic seemed to conjure up a more simplistic, although also popular, imaginary of the Arctic as being one of the 'last great wildernesses', and, as such, a place which should be protected or saved from human interference. Even in cases where people have been killed by polar bears (as happened on Svalbard in 2011, when a British 17-year-old, Horatio Chapple, was tragically mauled to death during an expedition), the response has been to debate whether polar bears protecting themselves from human intruders deserve more sympathy than people who lose their lives 'testing' themselves in the Arctic. More relevant to this book is that imagining the Arctic as a place that should be subject to minimal human interference—or even, as it was during Victorian times, as a sublime space to be feared (Spufford 1996)—often seems to obscure the idea that the Arctic might matter to contemporary Britain in other ways, relating, for example, to national security, energy security, environmental security, and economic prosperity,

This chapter responds by highlighting some of the ways in which Britain's long history of exploration, science, and trade in the Arctic have shaped ideas about British proximity to the Arctic, as well as how those ideas have evolved as the Arctic has variously been designated and used as a space for exploration, a colonial frontier, a resource province, a strategic theatre, and a scientific laboratory. Britain's contemporary interests in, and connections to, the Arctic are explored later in Chaps. 4 and 5.

Early Encounters

King Alfred the Great of Wessex was likely the first English sovereign to hear a detailed account of what must surely have seemed to be distant northern lands. In 890 AD, Alfred was visited by Orthere of Hålagoland, a Viking merchant and whaler who is believed to have lived above the

Arctic Circle, in what is now the Norwegian County of Troms. Orthere recounted to Alfred how he had travelled north and east along the coast of Northern Norway, eventually finding his way into what is now Russia, making it at least as far as the Kola Peninsula and the White Sea, where he had found large walrus colonies (Paine 2014).

Orthere then proposed a deal to Alfred: his knowledge of the icy Arctic seas and their many dangers, as well as ivory from the walrus colonies, in return for concessions on trading rights in Alfred's kingdom. With this offer, Orthere put, for the first time, a small part of what today constitutes the Arctic within England's reach. Alfred, however, turned down the deal, and, with it, a trade link to the Arctic. Like many others across Europe at the time, Alfred appeared to be sceptical of Orthere's claims about what lay in the North. He left it to the Vikings to give shape to the political landscape beyond the northernmost limits of known civilisation in Europe, in effect distancing himself and his kingdom from the Arctic (McCannon 2012; Paine 2014).

The persistence of the Viking Age until the eleventh century did, however, ensure that other connections between the British Isles and Scandinavia continued to exert a powerful draw to the North. Trade links between the English and Norwegians grew around the supply of, and demand for, fish, leading to the founding of Bergen in Norway in 1075. Competition between the English and the Hanseatics for Norwegian stockfish escalated until, by the fourteenth century, their fleets were fishing for themselves off the coast of Bergen. The Hanseatics, supported by the Kingdom of Norway, eventually gained the upper hand, driving the English fleets away. In response, the English turned to Icelandic waters, just south of the Arctic Circle. Such was the size of the English presence in Iceland that the fifteenth century is known among some Icelandic historians as the 'English Century' (Karlsson 2000). English fishing fleets were sailing further north than ever before, bringing about a new sense of proximity to the Arctic—a proximity which was certainly felt in Iceland, if not by the English themselves.

The Way of the North

As English fishing fleets sailed further into the North Atlantic and seas adjacent to the Arctic, they built up the knowledge, skills, experiences, and technologies that would later aid the rapid maritime expansion of the English Empire across the North Atlantic (ibid.). The Crown demanded trade and resources, and had heard of lucrative new markets to satisfy domestic appetites for silk, spices, and other goods in Asia. However, at that time, the Portuguese and Spanish Empires were in ascendance. Their dominance of the Atlantic Ocean was formalised in the Treaty of Tordesillas in 1494.[1] The English, the Dutch, and the French rejected the terms of the Treaty, but their own maritime power was still only nascent. Their access to the South Atlantic was blocked, so there was little prospect of establishing trade routes to Asia without Spanish or Portuguese interference.

In theory, the line demarcated by the Treaty of Tordesillas stretched from the far North to the far South. In practice, Portuguese and Spanish dominance was limited mostly to the South Atlantic. Henry VII, who wanted more maritime trade, looked north, and beyond the limits of the known world. In 1496, he commissioned John Cabot, a Genoese navigator, to seek a maritime passage to Asia through the North Atlantic. Cabot was to set sail from Bristol, which had become a hub of English expeditionary activity in the North Atlantic during the hunt for new fishing grounds. Following a precedent set by Richard II, Henry VII stipulated in the royal patent issued to Cabot that commerce resulting from any discoveries he might make must be conducted in England alone, while any territory in areas not yet reached by Spain would be claimed for England. As such, the patent represented an explicit rejection of the Treaty of Tordesillas (Mills 2003). It also represented what appears to have been the first attempt by an English ruler to expand his or her realm into the Arctic through trade, influence, and the claiming of 'virgin' territory.[2] Henry VII believed the Arctic to be within his reach.

Cabot's expedition was the first of many to set sail from the British Isles in the sixteenth and seventeenth centuries. The search for a northern maritime passage to Asia united the interests of the Crown, financiers, and adventurers in merchant cities such as London and Bristol. Even if a passage was not found, it was believed that the discovery of new territories and resources would fill the coffers of those, such as the Merchant Adventurers of London and the Drapers Company, who were prepared to invest in expeditions. Thus, the process of searching for a northern passage would be rendered financially productive regardless of the final outcome.

All three of Cabot's attempts to find a Northwest Passage failed. In 1521, when John Cabot's son Sebastian sought support for his own expedition to discover a Northwest Passage, he was turned away by Henry VIII, and other financiers, who presumably no longer saw any prospect of a financial return on such an expensive investment. However, across Europe, explorers and cartographers continued to speculate about what lay beyond the limits of the known world. In 1527, a merchant, Robert Thorne, who had been inspired by the work of Ancient Greeks, convinced Henry VIII that there was an Open Polar Sea in the Arctic, a theory which persisted among British elites into the nineteenth century.[3] Others believed routes might be found along the unchartered coasts of North America and northern Russia. What was evident in these debates was that owing to a persistent lack of knowledge about the Arctic, the region remained a blank canvas on to which early practitioners of geography and statecraft could project their theories. And they all agreed that, despite their ignorance, an Arctic passage to Asia must be within England's grasp.

Expeditions that foundered on encounters with ice certainly discouraged some explorers and their financial backers from further attempts. Yet because a failed expedition did not prove definitively that no passage existed, the possibility of eventually finding a passage remained a tantalising one. Religious conviction also played a part. The merchant and geographer Roger Barlow concluded in his *Briefe Summe of Geographie* (1541), which he presented to Henry VIII, that the 'waie of the northe' was something which God bestowed on the English alone (quoted in Wallis 1984: 453). John Davis, who made three attempts between 1585 and 1587 to discover a Northwest Passage, similarly believed that God had decreed no ocean could freeze, and that, consequently, the Arctic Ocean must be navigable 'for otherwise nature should be monstrous, and his creation wast' (quoted in Craciun 2010: 694). Despite periods of disinterest in the Arctic, wherein there were few if any attempts to send ships north, the English (i.e. the Crown, the merchants, the adventurers, and the scholars) were again and again drawn back by speculation and belief that the Arctic was theirs to overcome and conquer (Wallis 1984; Craciun 2010).

Mobilising the Arctic

In 1555, the Muscovy Company was issued a Royal Charter that gave it exclusive rights to all commercial activity in the North. The issuance of the Royal Charter followed an expedition led by Sir Hugh Willoughby, a former soldier turned explorer, to seek out a Northeast Passage. Willoughby was guided by Sebastian Cabot, two influential English geographers (John Dee and Richard Eden), and the Flemish cartographer Gerard Mercator (a famous proponent of Open Polar Sea Theory). Although much of the expedition ended in disaster (two ships were lost and their crews froze to death), several survivors, under the command of Richard Chancellor, returned, but only after making contact with the embryonic Russian state led by Ivan the Terrible. The Muscovy Company was subsequently created to build a trade route to the Muscovites, which would bring furs and other valuable resources from the Arctic to England.

In the 1590s, the Company began hunting whales and walruses, first in waters around Novaya Zemlya and Bear Island, and later off Spitsbergen (Israel 2002). As well as mobilising Arctic resources to be returned to England, the Muscovy Company claimed Spitsbergen and its waters as a sovereign possession of the Crown—an early attempt to push English borders into the Arctic, and, if seen by today's standards, establish the English as an Arctic nation. However, the English claim was contested by the French, Flemish, Danish-Norwegians, Spanish, and Dutch. Competition between those countries led eventually to the partitioning of Spitsbergen's harbours among the rival powers (Stonehouse 2007). The land itself remained without sovereignty, a *terra nullius*, or 'no man's land'.

The pattern of traders following in the wake of expeditions to seek out northern passages was repeated across the Atlantic Arctic. The ill-fated Cathay Company, which broke from the Muscovy Company, attempted to repeat the latter's success in North America. In the late 1570s, the Muscovy Company backed another attempt to find a Northwest Passage, this time led by the explorer Martin Frobisher. Frobisher returned claiming to have found a strait without end, as well as vast deposits of ore believed to contain gold. The Cathay Company was formed to enable Frobisher to return to North America for more ore. However, the ore that Frobisher brought back was later discovered to be worthless iron pyrites ('Fool's Gold') and the Cathay Company was bankrupted. Nevertheless, Frobisher's discoveries, which included several encounters with Inuit on Baffin Island, spurred further English interest in North America. Details of Frobisher's expeditions were published in three books prepared by Richard Hakluyt, an influential English geographer, in 1582, 1589, and 1598, volumes which have since been described as a 'prospectus for English enterprise in northern regions' (Wallis 1984: 454).

Hakluyt's work proved influential. It situated the quest for a Northwest Passage within a wider vision of English maritime expansion across the Atlantic. Following the defeat of the Spanish Armada in 1588, the English were becoming the leading maritime power in northern Europe, with North America at its mercy. Hakluyt encouraged the Crown to colonise North America as quickly as possible. English traders were being pushed out of the northeastern Arctic as Russia expanded its own presence in the region, and the influence of the Muscovy Company was waning. Hakluyt argued that by colonising North America, the English would be in a better position—owing to greater physical proximity—than their European rivals to seek out their Arctic prize. Following the advice of Hakluyt and others, the English went on to make significant territorial acquisitions north of the Arctic Circle. Along the way even more Arctic resources were mobilised to support the economy back home. In 1670, the Hudson's Bay Company received a Royal Charter from Charles II guaranteeing it control over the fur trade across much of British North America, including its 'Arctic' territories known as Rupert's Land.

The extent of English resource extraction in the Arctic was evident in the way whaling, sealing, and the North American fur trade became integral parts of the economies of several British cities in the centuries that followed. Bernard Stonehouse of the British Arctic Whaling Unit at the University of Hull has estimated that at least 35 ports (among them London, Hull, Liverpool, Exeter, Whitby, Newcastle, Sunderland, Lynn, Scarborough, Dartmouth, Dundee, Aberdeen, and Peterhead), hundreds of ships, and thousands of men and women were involved in the Arctic whaling industry in the eighteenth century. Whale oil, whale bone, and baleen became everyday commodities in the lives of British citizens, and brought light to British streets. Processing these and other resources provided work for thousands in factories producing lamp oil, machine lubricants, soaps, paints, dress hoops, and corsets, among many other things. The scale of activity powered further expansion of Arctic industry, driving shipbuilding and demand for ever more resources (Stonehouse 2007). As England/Britain's commercial and geopolitical reach expanded into the Arctic, the bounty returned became ever more prevalent in national life. By the close of the eighteenth century, Britain had become a major part of the Arctic, topographically and topologically, on account of its territorial gains in British North America, its persistent whaling interests around Greenland and Spitsbergen, and the imbrication of economic and social life with resources from the Arctic in many towns and cities across Britain.

The Heroic Age

By the nineteenth century, the British Empire had all but abandoned its search for a Northeast Passage. A regular flow of resources was being returned from the Arctic, strengthening the British hold on North America in particular. The motivation to be the first to discover the Northwest

Passage remained strong, although its character had changed since Tudor times. Then, the Crown, merchants, and adventurers had been seeking a new passage to the lucrative markets of Asia. Failing that, they turned their attention to the Arctic's valuable resources. In the eighteenth century, a Member of Parliament, Arthur Dobbs, complained that the Hudson's Bay Company had deliberately stopped the search for a Northwest Passage as it feared further discoveries along the coast of North America might break its trading monopoly.

Dobbs went on to persuade Parliament to offer a £20,000 reward to an expedition he was backing, if it led to the discovery of a Northwest Passage. The reward was written into legislation in 1745. In 1776, Parliament revised the legislation to allow the award to be granted to whoever was first to find any passage from the Atlantic Ocean to the Pacific Ocean running north of 52°N; £5000 would be awarded to the first expedition to reached 89°N (David 2000). In the early nineteenth century, Parliament replaced these rewards with a new scheme aimed at encouraging would-be explorers to seek out the Northwest Passage. The £20,000 reward for discovering the Northwest Passage was retained, but added to it were smaller rewards of £5000 for crossing 110°W and £10,000 for crossing 130°W north of the Arctic Circle. However, the availability of such awards suggests that the trade and resource monopolies already established in the Arctic had squeezed the opportunities for would-be merchants and adventurers to make their fortunes in the North. Meanwhile, the spread of the British Empire around the world had created greater opportunities elsewhere, especially in Africa and the Caribbean.

A different sort of inducement was now required, in the form of direct reward and recognition. The Arctic was to be rendered productive of something which, despite previous endeavours, still seemed altogether new—heroism. There was perhaps no one that embraced that more than John Barrow, the Second Secretary to the Admiralty, who turned the search for a Northwest Passage into a matter of national prestige, religious conviction, and moral worth. Barrow was convinced by the Open Polar Sea Theory and was in a position to put his beliefs to the test. Following the Napoleonic Wars (1799–1815), Barrow had a demobilisation problem. The numerous seamen who had been pressed into service were easy enough to turn loose. However, the naval officers, many of whom had political connections, were not so easy to remove from service, despite the fact that most of them had little to do, their pay had been halved, and there was little prospect of career advancement (Fleming 2001). Barrow had officers and ships to spare. In 1816, he argued that the British Empire could use this surplus to drive a new wave of geographical and hydrographical exploration which would further the scientific knowledge of the nation, and potentially provide a boon to national commerce following the recent wars in Europe (ibid.).

Barrow further believed it would be unwise to let other nations, with their own imperial ambitions, take on the challenge, lest it undermine Britain's own power and glory. His concerns were matched by a wider anxiety among the political elite about the moral worth of the British Empire at the beginning of the nineteenth century. The British Empire had become more than just a domineering, expansionist power. Its rulers believed the Empire as an expression of God's will. After the abolition of slavery, they had sought to recover the Empire's moral purpose by spreading 'civilisation' to the 'savages', a mission that both informed and justified its global reach. However, as the literary theorist and historian Adriana Craciun (2010) has shown, by resisting British sailors, the Arctic seemed to present an affront to that vision of the Empire's global reach and moral worth, much as it had to the God of Elizabethan explorers.

Barrow's ambitions were fuelled further by news from whalers sailing off the coast of Greenland that the sea ice on the edges of the Arctic Ocean was in retreat, and that a Northwest Passage might be opening. Writing later in 1846, John Barrow declared that to have not gone in search of a Northwest Passage at that time 'would have been little short of an act of national suicide' as it would have risked giving up the glory of conquering the Arctic to a maritime rival (Barrow 2011: 16). The changing Arctic environment had created new possibilities of access and connectivity that seemed to finally bring a Northwest Passage within reach of the British Empire.

Retreat from the Arctic

A succession of failed Arctic expeditions followed, culminating in the disastrous loss of Sir John Franklin's expedition in 1847. Following the disappearance of the *Erebus* and the *Terror*, along with Franklin and his crew, search and rescue missions were launched from both the Atlantic and the Pacific, as well as overland up through the northern reaches of British North America. All to no avail. At home, the disappearance of the Franklin expedition dominated public attention. Front page news regularly conjured up images of a sublime and monstrous Arctic where British explorers had been forsaken by God, turned to cannibalism, and met their doom (Spufford 1996).

The search for Franklin ended after missions led by John Rae and Leopold McClintock in 1854 and 1857 discovered relics, and later corpses and a cairn, that confirmed Franklin's death. Meanwhile, while searching for Franklin on another expedition, Robert McClure found proof that a Northwest Passage did exist, although ice made it impassable. With that discovery, together with the confirmation of Franklin's fate, the Admiralty, Parliament, and the Government were at a fork in the road. The search for the Northwest Passage had incurred great cost. To find that it was unnavigable seemed to confirm that the Arctic Ocean could not be assimilated into the global maritime trading network that underpinned the British Empire.[4] Although the Northwest Passage was closer than ever, the British lacked the means to traverse it in any useful or efficient fashion, rendering its physical location virtually meaningless.

In the latter half of the nineteenth century, Britain began its retreat from North America. In 1870, the territories of the Hudson's Bay Company were transferred to the newly establish dominion of Canada. In 1871, the first British North America Act was passed. However, uncertainty remained over exactly which territories, including those in the Arctic, the Hudson's Bay Company actually owned. With fears that the United States might seek to take advantage of the ambiguity to acquire additional territory in the North following its purchase of Alaska from Russia in 1867, the British Government decided to pass all of its North American Arctic territories to Canada (with the exception of Newfoundland), completing the transfer in 1880 (Smith 1961). As the British Empire unknowingly gave up the possibility that, one day, the future United Kingdom might be regarded as an Arctic state, for Canada and the United States, the opposite occurred—the territories they now controlled would ensure their future position as leading states in Arctic affairs.

By the end of the nineteenth century, the British Empire had largely withdrawn from the Arctic in topographical terms. Its only remaining territorial interests were in Spitsbergen, which was still considered *terra nullius*, and Newfoundland, which Canada eventually acquired in 1907. However, in topological terms, the picture was more complex. Whaling around Spitsbergen and Greenland continued into the early twentieth century. A British coal-mining company—the 'Spitsbergen Coal & Trading Company'—was opened in a mining settlement in Spitsbergen in 1904.

Yet, the persistence of British economic activity in the Arctic has attracted far less attention from historians than the so-called 'heroic expeditions'. Once the heroic age in the Arctic ended, topological connections between Britain and the Arctic, which had been more or less ongoing for centuries, were overshadowed by growing British interest in Antarctica at the end of the nineteenth century, and the emergence of Robert Scott and Ernest Shackleton as the new heroes of British polar exploration. The Government backed a series of expeditions to Antarctica which would eventually enable Britain to claim a piece of Antarctica for itself in 1908, and again in 1917.

Some have suggested that the switch occurred because British maritime power had reached its limits in the Arctic Ocean. Antarctica, on the other hand, despite being much more distant physically, was a continent (as opposed to a frozen ocean), and therefore easier to grasp, both imaginatively and physically (Craciun 2010). After all, it was the liminal qualities of ice that had enticed so many British merchants and explorers to attempt to sail across the Arctic Ocean, despite also being frustratingly opaque and challenging to navigate (Dodds 2018).

However, while further exploration of the Arctic was taken on by others such as the legendary Norwegian explorer Fridtjof Nansen, the British did not ignore the Arctic completely. The Royal Society and the Royal Geographical Society (RGS) persisted in reporting on and debating the discoveries that were being made in the Arctic. For example, both Fridtjof Nansen and Roald Amundsen, another famous Norwegian polar explorer, were invited to give lectures at the RGS about their Arctic expeditions. The accomplishments of others, such as Robert Peary, an American explorer who claimed to be the first to reach the geographic North Pole in 1909, were also deliberated upon. As the discoveries and achievements (especially 'firsts') mounted up, Clements Markham of the RGS concluded that the age of Arctic exploration was almost at an end (Markham 1902).

By the early twentieth century, there were fewer and fewer opportunities for British explorers to seek glory in the Arctic, even if they had found the support and resources needed to do so. However, the Arctic was already beginning to matter to Britain in another way. Awareness was growing in the scientific community that natural and physical events in the Arctic mattered in the mid-latitudes, and vice versa. Whereas just a few centuries earlier, Britain had been stretched geopolitically to encompass the Arctic, at the beginning of the twentieth century, scientists began to realise that regardless of geopolitical boundaries, the Arctic's reach could not be escaped. Again, this phenomenon is something which can only really be understood in topological as opposed to topographical terms. After all, the Arctic and the British Isles were no closer together physically. What *was* changing was that a growing number of connections between the Arctic and Britain were either being discovered or forged through natural (e.g. climate, weather), technological (e.g. the advent of the aeroplane), and industrial processes (e.g. the spread of pollution).

The Wider Arctic

The term 'Global Arctic' is gaining currency as scholars seek to understand the ways in which modern phenomena such as globalisation, transboundary pollution, and climate change have brought the Arctic into global webs of science, commerce, and security. However, the extent to which some of these webs, especially those relating to science and commerce, are 'modern' is questionable. Scientists continue to discover new connections between the Arctic and the rest of the world, but theories concerning the importance of the Arctic and Antarctic to natural processes—sometimes described as 'earth systems'—date back at least to the nineteenth century. Then, in Britain and elsewhere, the 'earth sciences' were a matter of imperial concern. Scientists and practitioners of statecraft were increasingly interested in understanding how magnetism and other meteorological phenomena might impact navigation and weather prediction around the world—knowledge that was of critical importance to the spread and maintenance of the British Empire, for instance. From 1882 to 1883, countries from across Europe and North America staged the International Polar Year (IPY), a major collaborative programme primarily studying the Arctic. Although Britain's contribution was relatively modest, it did partner with Canada to establish an observatory to monitor magnetic phenomena in the Arctic in Fort Rae, the site of a former Hudson's Bay Company trading post.

Between the First and Second World Wars, British interest in Arctic science and exploration increased again. In a paper entitled 'The Future of Polar Exploration' which he presented to the RGS in December 1920, Frank Debenham argued that the invention of aeroplanes that could fly longer distances during the First World War, and the possibility of turning the international cooperation and coordination that had won the war towards solving the problems of peacetime, meant that Britain was once again in a position to be a pioneer of Arctic exploration (Debenham 1921). Moreover, Debenham seemed to anticipate future interest in a 'Global Arctic'. For Debenham, scientific investigation of the Arctic was not just about the Arctic itself—it was about understanding where the Arctic (and indeed the Antarctic) fit with the

> world as a whole ... to realise that all parts [of the world] are interrelated in a most complicated way ... [and] that the polar regions do in all probability contain the key to world problems in science which may at any time make an enormous difference in practical affairs. (Ibid.: 191–2)

In other words, while the heroic age of exploration might be over, Debenham believed Britain should still care about the Arctic because what happened there contained so much potential to help resolve scientific problems of global importance, and, as Debenham also noted, who knew what commercial opportunities that might create for Britain as a result.

Shortly after Debenham presented his paper, he became the first director of the newly created Scott Polar Research Institute in Cambridge (SPRI). SPRI was to serve as an archive and knowledge centre for all scientific activity in the polar regions. Established within the University of Cambridge, it became one of the most important centres for polar research and education in the world, a status that it largely retains today, despite Britain no longer being an 'Arctic state'. Elsewhere, during the 1920s and 1930s, other British universities (including Oxford, Durham, Aberdeen, Imperial College, Birmingham, and Nottingham, among others) became actively involved in the scientific exploration of the Arctic. Expeditions left for Greenland, Spitsbergen, Iceland, and the Canadian Arctic. James Wordie of the RGS would eventually note that these expeditions 'marked a completely new phase of Arctic exploration from [Britain]' (Chetwode et al. 1939: 133). However, it ended abruptly with the outbreak of the Second World War.

A particularly interesting feature of the university-led expeditions was the way in which they fostered a scientific understanding of Britain and the Arctic as physically connected—or more accurately, as part of a continuum (see, for example, Longstaff 1929). Shared geological features, bird populations and ecosystems, as well as interrelated weather systems, all strengthened the scientific basis for seeing the British Isles and parts of the Arctic as possessing a shared materiality. That, in turn, drew attention to a kind of physical proximity not always evident on maps, something which has since come to resonate with contemporary debates about how the impacts of climate change in the Arctic affect Britain, how industrial pollution from Britain affects the Arctic, and whether Britain might be considered part of a 'Wider Arctic'. After all, for many scientists, the Arctic Circle at 66°N is a poor way to define where the Arctic begins and ends.[5]

The Arctic continued to attract British scientific attention throughout the rest of the twentieth century. As Debenham had argued, knowledge of the Arctic was not just important in its own right, but also because it contributed to scientific understanding of natural processes elsewhere in the world, including in Britain. As scientific interest in the Arctic expanded to encompass atmospheric sciences, oceanography, ice dynamics, the ionosphere and biological systems, interest also increased in understanding the teleconnections that enable the Arctic to affect, and be affected by, these systems on a global scale. Such knowledge could also be used to support commercial endeavours such as the attempt, in the 1920s and 1930s, to establish a trans-Arctic air route that would connect Britain and North America via Greenland, and later in efforts to mine the Arctic for oil, gas, and other mineral resources in places such as Alaska. It would also have strategic value as advances in weather forecasting, satellite communications, submarines, air transport, and ballistic missile technology during the Second World War and the Cold War were all reliant on a sound understanding of key physical processes. And it has been shown to be of further value still as concern in Britain (and internationally) has grown about climate change, and the spread of long-distance pollution, both of which impact and are impacted by what happens in the polar regions.

(S)he Who Controls the Arctic

In 1917, in the midst of the First World War, the Council of the RGS wrote to the British Secretary of State for Foreign Affairs warning of the threat that Germany would pose to Britain should it seize control of Spitsbergen. Specifically, the Council was concerned about Spitsbergen's 'relative proximity to the British isles ... only some sixty to seventy-two hours steaming from Scottish ports' (RGS 1918: 246–7). Although the response from the Foreign Secretary was underwhelming—he merely noted that the Government would keep the Council's concerns in mind—the letter was revealing of the way in which the strategic imperatives of the First World War led some to forge new connections between Britain and the Arctic that emphasised their strategic closeness. Elsewhere in the letter, the Council noted Britain's other interests in Spitsbergen, including in the steam coal deposits and in the nearby sea lines of communication which the Royal Navy was using to provide support to Britain's allies in northern Russia. To protect these interests, Britain pushed its front lines deeper into European Arctic, reaching as far as northwest Russia. The British presence in Arctic Russia was maintained until 1919 when the last British troops were withdrawn from Archangelsk and Murmansk.

Similar strategic concerns emerged during the Second World War, when Britain once again used sea lines of communications through Arctic waters, this time to support the Soviet Union and block Germany's access to the North Atlantic. Following advances in weather forecasting, the Arctic also became important for the siting of weather stations which provided critical information for the war effort further south, leading to a battle in the Arctic between Allied and Axis forces seeking to disrupt each other's operations (Liversidge 1960). As during the First World War, it was only by pushing its forces deeper into the region (and pushing German forces out) that Britain could safeguard a vital part of the Allied war effort.

During the Cold War, this strategic imaginary returned again, only now it encompassed most of the Arctic Ocean. For a third time, military forces provided the main link between Britain and, especially, the European Arctic—a connection once maintained by whalers. As British submarines patrolled beneath the Arctic ice cap, the British Isles became part of the so-called Greenland–Iceland–UK (GIUK) 'gap'. Britain was now a gatekeeper of air, surface, and subsurface travel between the North Atlantic and the European Arctic. Britain also took command of North Atlantic Treaty Organization's (NATO's) rapid reaction force—Allied Command Europe (ACE) Mobile Force (Land)—which was maintained at high readiness to deploy to the aid of Norway in the event of a Soviet invasion from the North.

Observing these developments, the Northern Waters and the Arctic Study Group (hereafter, Northern Waters Group), which from the 1970s to the early 1990s was part of the Scottish Branch of the Royal Institute of International Affairs, identified Britain's principal strategic interest in the Arctic as being related to the warmer maritime spaces of the North Atlantic and the European Arctic which bordered the Arctic Ocean. Clive Archer and David Scrivener, who led the Northern Waters Group's work, described these spaces as important because of the areas they joined, providing sea lines of communication between North America and Western Europe, as well as the approaches to the Arctic Ocean (Archer and Scrivener, 1986). The persistence of strategic vulnerabilities along NATO's 'Northern Flank' meant Britain, having the most powerful navy in Europe, had to maintain a forward defence presence in the Arctic as part of the broader NATO effort to contain the Soviet Union and keep the front line between East and West as far away from the North Atlantic as possible (Staveley 1988).[6]

Conclusion

British interest in the Arctic, then, did not end with the giving up of its Arctic possessions, or its retreat from Arctic exploration at the end of the nineteenth century. While the retreat of the British Empire's borders increased Britain's topographical distance from the Arctic, topologically Britain continued to maintain, discover, and forge new scientific, commercial, and military connections with the Arctic throughout the twentieth century. While these connections were partly produced through a sense of adventure, scientific curiosity, commercial opportunity, and strategic concern, they also to some extent demonstrated the Arctic's refusal to be ignored by Britain as new scientific, commercial, and strategic imperatives emerged in and centred on the Arctic and its influence on events in lower latitudes.

'Topography' and 'topology' offer two very different ways of thinking about Britain's proximity to the Arctic. While not mutually exclusive, preponderance towards using the former over the latter has had practical consequences for how Britain relates to the contemporary geopolitics of the region. That is most evident in the recent adoption of the phrase 'the Arctic's nearest neighbour' by successive British governments since 2010. That phrase simultaneously positions Britain as a non-Arctic state, while using the fact of topographical proximity as leverage for greater influence in the region—and, crucially, more influence than other non-Arctic states, which might have other sources of influence, finance, for example.

The problem with this approach it that it overshadows the far more extensive topology that Britain continues to share with the Arctic. Historical events, such as the transfer of British Arctic territories in North America to Canada, or the purchase of Alaska by the United States, demonstrate the arbitrariness of using topographical logics to define the 'Arctic-ness' of a state, or, indeed, identity in general. Between 1864 and 1944 alone, the number of Arctic states changed at least seven times before ending up at the current eight (Mazo 2015). While the latest constellation has proven more durable (lasting more than seven decades so far) it is still vulnerable to further change, for example, if Greenland and the Faroe Islands ever become independent from Denmark.

The British claim to be the 'Arctic's nearest neighbour' is similarly threatened by the spectre of Scotland one day becoming an independent nation. Yet were Scotland to become independent, few would seriously question whether the rest of Britain's interest in the Arctic should be at all diminished, or that Scotland should have a greater role than the rest of Britain in Arctic affairs, and that is because of the extent of the topology Britain would continue to share with the Arctic (through science, commerce, and military strategy as well as concerns about climate change, pollution, and conservation). The contemporary topology is explored in the following chapters, but the impetus for taking this topology seriously is to be found in the history of Britain's relationship with the Arctic, some key elements of which have been outlined in this chapter. That history repeatedly emphasises the point that the importance of the Arctic to Britain did not begin and end with the acquisition and release of lands north of 66° latitude. Even when Britain's priorities have seemingly been in other parts of the world, the Arctic continued to attract attention from a wide range of actors who rendered the region productive in scientific, commercial, strategic, and even moral terms.

That Britain's historical connections to the Arctic appear to have been largely forgotten, or at least neglected, is perhaps unsurprising given that one of the most enduring legacies of the British Empire is that Britain has, and continues to adopt, a global perspective when it comes to its foreign policies and overseas interests. Over the period when Britain might have imagined itself as an 'Arctic state' topographically (c. 1670–1880), it was also expanding its reach into the Caribbean, West Africa, India, and China. With such a vast empire, it is perhaps no wonder that an 'Arctic' identity was never established in the way that it has come to grip the foreign policy establishments of many of today's Arctic states.[7] As the next chapter shows, 'Arctic-ness' itself is a relatively novel phenomenon—born of a desire to support Arctic nation-building projects and clearly delimit the principal actors in regional affairs, something which even non-Arctic citizens of Arctic states have struggled to identify with (Keskitalo 2004).

Although successive British governments since 2010 have tried to stake out a 'near-Arctic' identity (along with China), there is little evidence to suggest that it has resonated with the rest of the nation, especially when British foreign policy has tended to be tracked and measured by scholars, practitioners, and the media against its relationships with the United States, European Union (EU), and NATO—and more recently China—as well as its ability to maintain security at home through overseas interventions, for example, in Iraq, Afghanistan, and Syria. In the absence of an Arctic topography, it is perhaps only by drawing attention to the extensive topology that Britain continues to share with the Arctic (manifesting, for example, as concern about climate change and national security, interest in new economic opportunities, and even literature and exhibitions that offer reappraisals of how polar exploration shaped British history and culture) that the contemporary importance of the Arctic to Britain becomes apparent.

The Arctic's importance to Britain has shifted over time, and, as a consequence, so too has its topological proximity. This in turn provides the basis for understanding Britain's relationship with the Arctic as something which is dynamic, shaped by connections between different actors, sites, knowledges, and practices, rather than the ahistorical topographical facts. Moreover, it provides the basis for questioning why it is that in the early twenty-first century, and animated by growing global interest in, and globalisation of, the Arctic, northern latitudes once again seem to loom large in the minds of at least some British scientists, defence planners, industries, and civil society, and how it is they are once again trying to render the Arctic into something which is scientifically, commercially, strategically, and morally productive.

Notes

1. The treaty divided possession of the 'New World' between the two empires by drawing a line in the Atlantic Ocean about 370 leagues west of the Portuguese-controlled Cape Verde Islands. All lands east of that line (approximately 46°37′W) were claimed by Portugal. All lands west of the line were claimed by Spain.
2. Notwithstanding the claims indigenous peoples already living in the North might have to Arctic lands and seas.
3. The theory was based on a number of 'facts': that ice only forms close to land; that ice can only form in freshwater; and that any ice that did form could not withstand the heat of the summer sun. Accordingly, Thorne was reported to have claimed that 'there is no land uninhabitable, nor Sea innavigable' (Mills 2003: 484).
4. Meanwhile, on land, the Hudson's Bay Company's attempts to expand the British fur trade deeper into Inuit lands had also been frustration (Stuhl 2017).
5. For more on different ways of defining the Arctic see Nuttall (2005).
6. The northern waters also mattered for their economic value as a source of living and non-living resources feeding the economies of North America and Western Europe. The so-called 'Cod Wars' between Britain and Iceland over fishing rights in the 1950s and 1970s had demonstrated the risks that economic competition between member states posed to NATO unity (Jónsson 1982).
7. Interestingly, it is the United States, another state with a global outlook, that has been the most resistant to embracing a specifically 'Arctic' identity.

References

Archer, Clive, and David Scrivener. 1986. Introduction. In *Northern Waters*, ed. Clive Archer and David Scrivener, 1–10. Totowa: Barnes & Noble.

Barrow, John. 2011. *Voyages of Discovery and Research Within the Arctic Regions: From the Year 1818 to the Present Time*. Cambridge: Cambridge University Press.

Chetwode, Philip, R. Scott Russell, J.M. Wordie, J.N. Jennings, N.E. Odell, and Alexander King. 1939. The Imperial College Expedition to Jan Mayen Island: Discussion. *The Geographical Journal* 94: 131–134.

Craciun, Adriana. 2010. The Frozen Ocean. *PMLA* 125: 693–702.

David, Robert. 2000. *The Arctic in the British Imagination 1818–1914*. Manchester: Manchester University Press.

Debenham, Frank. 1921. The Future of Polar Exploration. *The Geographical Journal* 57: 182–200.

Deleuze, Gilles, and Félix Guattari. 2004. *A Thousand Plateaus*. London: Continuum.

Dodds, Klaus. 2018. *Ice*. London: Reaktion Books.

Doel, Marcus. 1996. A Hundred Thousand Lines of Flight: A Machinic Introduction to the Nomad Thought and Scrumpled Geography of Gilles Deleuze and Félix Guattari. *Environment and Planning D: Society and Space* 14: 421–439.

Elden, Stuart. 2005. Missing the Point: Globalization, Deterritorialization and the Space of the World. *Transactions of the Institute of British Geographers* 30: 3–19.

English, John. 2013. *Ice and Water: Politics, Peoples and the Arctic Council*. Toronto: Allen Lane.

Fleming, Fergus. 2001. *Barrow's Boys*. London: Granta Books.

Israel, Jonathan. 2002. *Dutch Primacy in World Trade: 1585–1740*. Oxford: Clarendon.

Jónsson, Hannes. 1982. *Friends in Conflict: The Anglo-Icelandic Cod Wars and the Law of the Sea*. London: Hurst.

Karlsson, Gunnar. 2000. *Iceland's 1100 Years: The History of a Marginal Society*. London: Hurst & Company.

Keskitalo, Ekaterina. 2004. *Negotiating the Arctic: The Construction of an International Region*. London: Routledge.

Liversidge, Douglas. 1960. *The Third Front*. London: Souvenir Press.

Longstaff, Tom. 1929. The Oxford University Expedition to Greenland, 1928. *The Geographical Journal* 74: 61–69.

Markham, Clements. 1902. Arctic Problems. *The Geographical Journal* 20: 481–484.

Mazo, Jeffrey. 2015. Showing the Flag. *Survival* 57: 241–252.

McCannon, John. 2012. *A History of the Arctic*. London: Reaktion Books.

Mills, William James. 2003. *Exploring Polar Frontiers: A Historical Encyclopedia: Volume 1, A–L*. Santa Barbara: ABC-CLIO, Inc.
Nuttall, Mark, ed. 2005. *Encyclopedia of the Arctic: Volumes 1, 2 and 3 A–Z*. Abingdon: Routledge.
Paine, Lincoln. 2014. *The Sea and Civilisation: A Maritime History of the World*. London: Atlantic Books.
RGS. 1918. British Interests in Spitsbergen. *The Geographical Journal* 51: 245–249.
Smith, Gordon. 1961. The Transfer of Arctic Territories from Great Britain to Canada in 1880, and Some Related Matters, as Seen in Official Correspondence. *Arctic* 14: 53–73.
Spufford, Francis. 1996. *I May Be Some Time: Ice and the English Imagination*. London: Faber and Faber Limited.
Staveley, William. 1988. An Overview of British Defence Policy in the North. In *Britain and NATO's Northern Flank*, ed. Geoffrey Till, 65–73. London: Macmillan Press.
Stonehouse, Bernard. 2007. British Arctic Whaling: An Overview. *British Arctic Whaling*. Accessed December 9, 2016. https://www.hull.ac.uk/baw/overview/overview.htm
Stuhl, Andrew. 2017. *Unfreezing the Arctic*. Chicago: University of Chicago Press.
Taylor, Peter. 1994. The State as Container: Territoriality in the Modern World-system. *Progress in Human Geography* 18: 151–162.
Wallis, Helen. 1984. England's Search for the Northern Passages in the Sixteenth and Early Seventeenth Centuries. *Arctic* 37: 453–472.

CHAPTER 3

The Circumpolar Arctic

Abstract Britain remains at risk of being peripheralised in Arctic affairs because of the new geopolitical order that emerged in the region after the Cold War. That geopolitical order was defined by a process of circumpolarisation, through which the eight Arctic states sought to establish their primacy over the region, even in areas beyond their national jurisdiction. This chapter investigates the history of circumpolarisation and how, despite various tensions, it has become central to shaping relations between the Arctic states, and between the Arctic states and the rest of the world. It further argues that if Britain and others are to challenge circumpolarisation from the outside they need to invest more in building up their connections with the Arctic.

Keywords Circumpolarisation • Peripheralisation • Topology • Topography • Global Arctic • Arctic geopolitics

Britain's proximity to the Arctic is determined by more than the extent of Britain's historical interest in the region. Chapter 2 described some of the ways in which Britain and the Arctic, imagined as discreet geographical containers, have been folded together or 'scrumpled' through the connections they share. Those connections have been forged and sustained over the past four centuries or so through networks of knowledges, sites, actors and practices related to exploration, trade, science, natural phenomena,

© The Author(s) 2018

D. Depledge, *Britain and the Arctic*,
https://doi.org/10.1007/978-3-319-69293-7_3

and strategic thinking, and draw attention to a topological proximity that is hidden from view on topographical maps.

But the possibility of *connecting* Britain and the Arctic also implies the possibility of *disconnection*—of disrupting the topologies that produce *proximity* and instead produce *distance*, as happened, for example, when Britain retreated from North America and the Northwest Passage in the latter half of the nineteenth century.[1] Over the last century in particular, connections between Britain and the Arctic have also been disrupted by interventions from the A8 that have sought to position Britain (and other 'non-Arctic' states) outside a 'Circumpolar Arctic' and its related structures.

For instance, on 19 September 1996, representatives from the Governments of Canada, Denmark, Finland, Iceland, Norway, Russia, Sweden, and the United States jointly issued the 'Declaration on the Establishment of the Arctic Council'. The Arctic Council was to be a high-level intergovernmental forum to

> provide a means of promoting cooperation, coordination and interaction among the Arctic states, with the involvement of the Arctic indigenous communities and other Arctic inhabitants on common Arctic issues [excluding security], in particular issues of sustainable development and environmental protection in the Arctic. (Ottawa Declaration 1996: 2)

The declaration consolidated the emergence in the early 1990s of an 'Arctic Eight' (A8): the international grouping which had started to coalesce during the last years of the Cold War. In doing so it institutionalised the idea that 'circumpolarity' should be the central ordering principle of Arctic geopolitics. In other words, participation and primacy in Arctic affairs should be defined by the Arctic Circle.

Circumpolarisation

The earliest geopolitical antecedent of the circumpolar approach to Arctic geopolitics was the Agreement on the Conservation of Polar Bears signed by Canada, Denmark, Norway, the Soviet Union, and the United States in 1973 (Oslo Agreement 1973). However, explorers and scientists, including British ones, had long imagined the Arctic as a discreet part of the world, providing the basis for initiatives such as the International Polar Year (IPY) from 1882 to 1883. Ultimately though, to think of 'circumpolarisation' as a process—through which a 'Circumpolar Arctic' region is constituted as a geopolitical container to act in/from—is to recognise both that the Arctic has not always been defined, organised, and administered in circumpolar terms and that the Arctic itself remains geographically stretchable (Kristoffersen 2014).

In the nineteenth century, natural scientists working on the polar regions—many of them out of places such as the Royal Society in London—recognised that the high latitudes shared many physical characteristics and phenomena, such as meteorology, magnetism, ocean currents, and the movement of ice. Further, they found those phenomena could only be studied by regarding the Arctic as a whole (although like their political counterparts, they also could not agree on where the Arctic began and ended), which no one nation could do alone without claiming sovereignty over the entire Arctic. That in turn provided the basis for 12 countries to stage the first IPY. Fifty years later, the international community staged a second IPY from 1932 to 1933, which reaffirmed the importance of a holistic and cross-border approach to scientific studies of natural phenomena in the polar regions.[2]

Up until the twentieth century, in the social and especially the political sciences, holistic framings of the polar regions were far less prominent. And over the twentieth century the geopolitical 'circumpolarisation' of the Arctic proceeded along a very different trajectory compared to the Antarctic. As the geopolitics scholar Klaus Dodds (2002) and others have argued, Antarctica is of course a continent rather than an ocean, and as such was more vulnerable to, and indeed became subjected to, several territorial claims between 1908 and 1942. International interest in Antarctica increased further during the Second World War. In 1948, with geopolitical tensions rising among claimant states (particularly between Britain, Chile, and Argentina), the United States floated a proposal for Antarctica to be brought under an international trusteeship—implicitly, if not explicitly, framing Antarctica in circumpolar terms. Although the initial proposal was blocked, it did set the Antarctic claimant states, together with the United States and Soviet Union, on a path which eventually led to the negotiation of the Antarctic Treaty. Signed in 1959 (entering into force in 1961), the treaty effectively regulated the terms of human activity on all land and ice shelves below 60°S.

In the Arctic, another decade passed before the A8 considered 'circumpolarity' as a basis for organising Arctic affairs.[3] Unlike Antarctica, the Arctic is mostly an ice-covered ocean surrounded by continents. Land claims above the Arctic Circle were largely settled by the mid-twentieth century. With little physical basis for a circumpolar 'Arctic Treaty', modelled on the Antarctic Treaty,[4] the circumpolarisation of the Arctic was instead driven by growing awareness among the A8 that despite geographic variation across the region they faced several common challenges in the high latitudes. For example, the aforementioned Oslo Agreement in 1973 was specifically negotiated to address the need for pan-Arctic cooperation among the Arctic Ocean littoral states on the issue of polar bear conservation. However, in subsequent years, further progress on pan-Arctic cooperation was difficult to achieve, largely because—with the exception of the Oslo Agreement—the Soviet Union maintained its long-standing resistance to using 'circumpolarity' as an organising principle for international cooperation, preferring bilateral arrangements instead (Stokke 1990).

Despite a lack of progress on circumpolar cooperation in the decades that followed, the precedent of organising international cooperation in the Arctic on circumpolar terms, established by the Oslo Agreement, was to persist throughout the rest of the twentieth century. That was facilitated by the apparent elasticity with which the A8 applied 'circumpolarity' to the Arctic. 'Circumpolarity', as established by the Polar Bear Agreement, was a limited one, covering only the Arctic Ocean littoral states (Canada, Denmark, Norway, the Soviet Union, and the United States, or 'A5'). That was a condition of Soviet participation in the treaty (Stokke 1990). However, in 1987, as part of a much broader 'New Thinking' on foreign policy, the Soviet Union started to relax its position on who/what constituted the Circumpolar Arctic. Speaking in Murmansk, the Soviet Premier Mikhail Gorbachev announced that

> the Soviet Union is now in favour of a radical lowering of the level of military confrontation in the region. Let the North of the globe, the Arctic, become a zone of peace. Let the North Pole be a pole of peace. (Gorbachev 1987: 4)

More broadly, Gorbachev wanted to see increased scientific cooperation among the 'rim' and 'sub-Arctic'[5] states, which to Gorbachev meant the A5 plus Finland, Sweden, and Iceland: effectively, the Circumpolar Arctic as we know it today.

Around the same time, the A8 started negotiating several 'circumpolar' agreements. The first was the International Arctic Science Committee (IASC, 1987–1990). Although originally established as a non-governmental scientific organisation to encourage and facilitate international cooperation in Arctic research (along similar lines to the Scientific Committee on Antarctic Research, or SCAR), government interest in the body gave it a distinct shape.[6] The Soviet Union was initially against opening up the IASC to non-Arctic states. Canada was prepared to allow non-Arctic states to participate but not on an equal basis with Arctic states (Smieszek 2015).

The United States, meanwhile, objected to the idea that the 'circumpolarity' should grant the Arctic states any primacy in decision-making about Arctic science. Washington's position was that larger forums were needed to deal with Arctic issues because so many problems and solutions originated from beyond the region. Britain and other non-Arctic states, unsurprisingly, also protested against attempts to limit their role in the new body. In the end, the A8 agreed that the IASC would comprise a policy- and decision-making Council including the A8 (Gorbachev's distinction between 'littoral' and 'sub-Arctic' states was dropped) and other states with significant Arctic research interests, as well as a Regional Board that would address the specific 'circumpolar' interests of the Arctic states, which included ensuring that the IASC's activities did not conflict with those interests (Smieszek 2015). Even in the Council, though, circumpolarity was implicit: non-Arctic states, *contra* Arctic states, had to maintain a significant scientific presence in the region if they were to retain their seat on the Council. Subsequent negotiations during the Rovaniemi Process (1989–1991), which produced the Arctic Environmental Protection Strategy, or AEPS (1991), and, later, the Arctic Council (1996) only reinforced the new circumpolar precedent that was established by the IASC negotiations.

Circumpolarity and the Arctic Council

The principle of circumpolarity became an important instrument for shaping the conduct of Arctic and non-Arctic actors. That happened despite the fact that the use of 'circumpolarity' as an organising principle is limited to just a few institutions, primarily the IASC, the Arctic Council, and associated forums (such as the Arctic Economic Council). Arguably the most important mechanism of Arctic governance is the United Nations (UN) Convention on the Law of the Sea (UNCLOS) that applies globally and that more than 160 countries are party to. UNCLOS provides the legal basis for most issues pertaining to the Arctic Ocean and to the Arctic states' continental shelves. The A5 also reaffirmed the centrality of the international law of the sea to Arctic Ocean governance in 2008 (Ilulissat Declaration 2008). Related to UNCLOS, the International Maritime

Organisation (IMO), another UN organ, has global responsibility for the safety and security of shipping and the prevention of marine pollution by ships, leading to the creation of instruments such as the Polar Code which came into effect in 2017. Arctic affairs are also shaped formally by regional and bilateral cooperative frameworks, which include organisations such as the Nordic Council, the West Nordic Council, the Barents Euro-Arctic Council, and even the European Union (EU) and North Atlantic Treaty Organization (NATO). These institutions effectively challenge the principle of circumpolar organisation by mobilising alternative constellations of states and other actors, some Arctic, some not, in arrangements that only cover parts of the Arctic (Young 2005; Stokke 2011).[7]

Unsurprisingly, Canada and Russia, as those Arctic countries most supportive of circumpolarity, have resisted suggestions that the EU and NATO—as the largest of those alternative organisations—should become more involved in Arctic affairs. More generally, circumpolarity has become a way for the Arctic states to organise and resist attempts by states and other actors from beyond the region to impose their own ideas and practices about how Arctic affairs should be governed. In similar fashion, indigenous peoples' organisations, such as the Inuit Circumpolar Council, have also sought to use circumpolarity as a basis for resisting what they regard as outside attempts to intervene in the affairs of their people.

The creation of the Arctic Council in particular has helped ensure that circumpolarity has remained the central organising principle in Arctic affairs despite only being one part of an Arctic governance 'mosaic' (Young 2005; Depledge and Dodds 2017). The Arctic Council's own emergence as a focal point in Arctic affairs largely coincided with the publication of the Arctic Council's *Arctic Climate Impact Assessment* (ACIA) in 2004 (Graczyk and Koivurova 2014). ACIA brought worldwide attention to the disproportionate impacts that anthropogenic climate change had already had on the Arctic, with worse to come in the future. It also drew attention to the global implications of Arctic warming, including, for example, global sea level rises, changes in ecosystems and vegetation cover, and additional amplification of the drivers of global warming (ACIA 2004). The findings were taken up by the UN's Intergovernmental Panel on Climate Change (IPCC) and also helped to raise the profile of the Arctic in the UN Framework Convention on Climate Change (UNFCCC).

ACIA attracted additional international interest in how environmental changes were affecting the Arctic's resource and maritime potential, including commercial possibilities related to shipping, mineral development, and fisheries (Ibid.). Subsequent international interest in the Arctic was therefore also motivated by emerging opportunities—highlighted, for example, by the United States Geological Survey (USGS) assessment of the Arctic's oil and gas resources in 2008—to capitalise on the impacts of climate change in the Arctic. However, then, as now, the Arctic was not a 'Wild West' and new commercial opportunities are largely limited to the existing national jurisdictions of the Arctic states. As a result, the A8 have emerged as the final political arbiters of any Arctic economic boom, at least for the foreseeable future.[8]

Even so, while the Arctic has a long history of being imagined as a verdant north, and more recently, as a $1 trillion ocean,[9] rendering it commercially productive remains a daunting task given much of the region's extreme environment, remoteness, low population, lack of material infrastructure and associated capital, and the sheer cost of operating there. There are also important considerations related to the protection of the environment and conservation of ecosystems, obligations that all of the Arctic states have committed themselves to, albeit more or less rigorously in terms of implementation, monitoring, and enforcement. That has heightened demand for the A8 to cooperate, using the Arctic Council in particular—and to a lesser extent, the more limited meetings of the A5—as a vehicle for addressing issues such as search and rescue, oil spill response, and scientific cooperation. They have also continued to use the Arctic Council to produce 'circumpolar' assessments on issues ranging from shipping to biodiversity to resilience (AMSA 2009; ABA 2013; ARR 2016). As the Icelandic historian Valur Ingimundarson (2014) has noted, the convergence of the A8 around their common circumpolar geography has been a key driver of the development of new structures of Circumpolar Arctic governance.

Much of the work being done to address the science of climate change and to agree terms promoting safe and sustainable commercial activity has been taking place under the auspices of the Arctic Council. The EU and non-Arctic states (as well as non-Arctic non-governmental organisations [NGOs]) have responded by pinpointing the Arctic Council as the primary forum for Arctic issues, and by increasing their engagement (Graczyk and Koivurova 2014). Over the past decade, there has been a surge of applications from non-Arctic states, NGOs, and the EU, to become Observers to the Arctic Council. In 1998, there were just four Observer states: Britain, the Netherlands, Germany, and Poland, all of which were also present at the signing ceremony for the Ottawa Declaration in 1996. They were joined by France in 2000.

All five countries were established actors in the Arctic scientific community and recognised for their commitment to dealing with Arctic environmental pollution, much of which originated—and continues to originate—from south of the Arctic Circle. They were also close allies of most of the Arctic states. Moreover, their support for the earlier initiatives to promote circumpolar cooperation among the A8 also suggests that while they might have wanted greater influence in Arctic affairs, they were prepared to accept the principle of 'circumpolarity'—that is the primacy of the A8—as the price of admission (Graczyk and Koivurova 2014; Dodds 2015).

After the Ottawa Declaration in 1996, the category of 'Arctic states' was more or less taken for granted, as was the category of 'Permanent Participants', which gives several indigenous peoples organisation seats at the negotiating table that are denied to Observers, and thus establishes their primacy over non-Arctic actors as well. But in 2002, anxiety started to become apparent among the Arctic states about whether an ever-greater influx of Observers might damage the Arctic Council, as well as pose an implicit threat to the principle of circumpolarity (Dodds 2015).

In 2006, Spain became the sixth permanent Observer to the Arctic Council. China and Italy expressed their interest in becoming permanent Observers in 2007 and started attending meetings on an ad hoc basis. That meant applying for permission from the A8 to attend each meeting. However, this picture of steadily growing interest in the Arctic was distorted by worldwide hype about a scientific expedition led by New York's famed 'The Explorer's Club', during which a Russian flag was planted on the sea bed directly beneath the North Pole in August 2007.[10] Peter Mackay, the Canadian Foreign Minister (2006–2007), in line with his Government's hard-line position on Canadian Arctic sovereignty, reacted strongly, famously declaring: 'This isn't the fifteenth century. You can't go around the world and just plant flags and say, "we're claiming this territory"' (Reuters 2007). Artur Chilingarov, one of the expedition leaders and reportedly a close friend of Russian President Vladimir Putin, had done little to help matters when he declared just before the expedition that 'the Arctic is ours and we should demonstrate our presence'. Other high-level Russian officials, meanwhile, were at pains to point out that the aim of the expedition was to gather scientific data to prove that the Russian continental shelf extended to the North Pole (BBC 2007a; see also Foxall 2014). Nevertheless, additional commentary in the Western media fixated on the idea that a new Cold War was beginning in the Arctic (Borgerson 2007; Carter 2007; Lucas 2008; see also Dodds 2008).[11]

Circumpolar Tension

In March 2008, the European Commission published a paper on the implications of climate change for international security. The paper referenced the Russian flag at the North Pole and went on to warn that rapid melting of the Arctic ice cap was 'changing the geo-strategic dynamics of the region with consequences for international stability' and that there was 'an increasing need to address the growing debate over territorial claims and access to new trade routes' (European Commission 2008: 8). Shortly afterwards, an article published by *Foreign Affairs*, a widely read foreign policy magazine, helped generate further media hype about the 'new scramble for territory and resources' under way in the Arctic (Borgerson 2008).[12] In August 2008, the A5 responded by gathering in Ilulissat, Greenland, at the invitation of the Danish Foreign Ministry, to hold high-level political discussions with the aim of ensuring a peaceful solution would be found to any potential conflicts over territory in the Arctic (Jacobsen 2016). The resultant Ilulissat Declaration stated that 'by virtue of their sovereignty, sovereign rights and jurisdiction in large areas of the Arctic Ocean, the five coastal states are in a unique position to address [emerging] possibilities and challenges', and, furthermore, that 'we ... see no need to develop a new comprehensive legal regime to govern the Arctic Ocean' (Ilulissat Declaration 2008: 1–2).

In making that Declaration, the A5 were reassuring each other and the rest of the world that, despite the rapid changes occurring in the Arctic, the status quo (read: the primacy of the Arctic states) remained intact (Steinberg et al. 2015). Yet arguably the most significant aspect of the Ilulissat Declaration was the exclusion of Finland, Sweden, and Iceland. Their exclusion suggested that the A5 felt that the Circumpolar Arctic had perhaps become too elastic, and, in the face of growing international encroachment, threatened to lose all meaning. Finland, Sweden, and Iceland were aggrieved, as were the Permanent Participants who were not invited either. When the A5 convened again in Ottawa in 2010, some were concerned that the split between the A5 and A3 would undermine the Arctic Council, at least until Hilary Clinton, the United States Secretary of State (2009–2013), walked out of the meeting early after criticising Canada for excluding other members of the Arctic Council (Woods 2010). Clinton had once again stretched Arctic circumpolarity to include the A3 and Permanent Participants.

Having flirted with discord and division, following the meeting in Ottawa the A8 refocused their attention on making common circumpolar cause around the need to strengthen the Arctic Council (Ingimundarson 2014; Dodds 2015). Their efforts proceeded along two paths. The first was to negotiate a series of multilateral treaties for the Arctic, under the auspices of the Arctic Council, to which only the eight Arctic states could be signatories. These treaties provided for circumpolar cooperation on Search and Rescue (2011), Marine Oil Pollution, Preparedness and Response (2013), and Arctic Science Cooperation (2017). By excluding from the treaties those states that do not hold territory above the Arctic Circle, the A8 were once again able to reaffirm their primacy over non-Arctic states.

The second was to tackle the question of what to do with the applications that were mounting up from non-Arctic states that wanted to be accredited as Observers to the Arctic Council. Following China's and Italy's expressions of interest in 2007, the EU and South Korea added their own in 2008, Japan in 2009, and Singapore in 2011. Practically speaking, the only difference that becoming an accredited Observer would make was that they would no longer have to apply for permission to attend meetings of the Arctic Council. However, symbolically, the granting of accredited Observer status was also seen as a tacit acknowledgement, by the A8, of a non-Arctic actor's legitimate Arctic interests (Stepień and Koivurova 2015: 35).

Between 2009 and 2013 the A8 deferred all Observer applications on the grounds that the Arctic Council needed to develop a new set of guidelines for Observer applications (Arctic Council 2013). The A8 went further still by developing rules governing the conduct of Observers at the Arctic Council. That included existing Observers such as Britain. The A8 was concerned that the international community was starting to encroach too much on their primacy in the management of Arctic affairs (Graczyk and Koivurova 2014). That was at least in part because, in 2008, the EU had approved a ban on the trade in commercial seal products, which Canada argued threatened the livelihoods of indigenous people in the Arctic. The European Parliament caused further consternation in 2008 when it called for 'the adoption of an international treaty for the protection of the Arctic, having as its inspiration the Antarctic Treaty' (European Parliament 2008).[13] Furthermore, in 2010, Chinese Rear Admiral Yin Zhuo was quoted by the official China News Service as saying that

> the Arctic belongs to all the people around the world as no nation has sovereignty over it… The current scramble for the sovereignty of the Arctic among some nations has encroached on many other countries' interests. (Chang 2010)

Both the European Parliament and the Chinese Admiral seemed to be arguing that the Arctic states themselves were somehow failing in their responsibilities and that others might have to step in. This was despite the fact that, as the A8 and Permanent Participants were quick to point out, environmental threats such as pollution and climate change are largely a consequence of excessive industrial activities in places far away from the Arctic, and are issues which the rest of the world already has the power to address without coming to the Arctic Council. Canadian Foreign Minister (2008–2011) Lawrence Cannon's accusation in 2009 that the EU lacked 'the required sensitivity' needed to be recognised as a permanent Observer to the Arctic Council was seen as a warning to all, including existing Observers such as Britain, that failure to respect the primacy of the Arctic states in Arctic affairs could see them excluded from the Arctic Council (CBC 2009).

The Arctic Council's decision to review the role of Observers had several consequences. The first was to raise up new Observers to the level of existing Observers. The previous distinction between permanent and ad hoc Observers was dropped to dispel any notion that the existing Observers were somehow more privileged. Existing Observers were also asked to provide the Arctic Council with detailed information demonstrating their value to the Arctic Council on the basis of what they had contributed over the years. The new rules on Observers further levelled the playing field by setting out that the status of all Observers would be subject to regular review. According to a Foreign Office official who was there at the time, that was something that Britain only grudgingly accepted as it was felt that such a tick-box exercise would produce a greater focus on quantity over quality of engagement, as well as a lack of flexibility over what counted as participation.[14]

The second was to force the new Observers to publicly affirm the sovereign authority of the Arctic states and demonstrate their willingness to support the work of the Arctic Council—just as existing Observers had done—so that there could be no doubt about the ongoing primacy of the Arctic states in Arctic affairs. By accepting these terms, both old and new Observers were effectively accepting 'circumpolarity' should be the basic organising principle of Arctic affairs or, in other words, that those countries lying below the Arctic Circle in topographical terms, and their interests, are secondary to those lying above it. As a consequence, Observers now ran the risk that, should they take any actions that could be interpreted as being counter to the interests of the A8, they might be expelled from the Arctic Council. Principal among the proscribed acts would be any attempt to challenge the primacy of the A8 and the Arctic Council in Arctic affairs. Thus, even though the Arctic Council remains a fairly limited organisation in practical terms (its mandate is limited to issues concerning 'circumpolar' sustainable development and environmental protection; and it only addresses issues on which there is consensus among the A8—see Ingimundarson 2014), and even though the A8 themselves regularly speak of the heterogeneity of the Arctic, symbolically it has come to enable the A8 states to perform a kind of exclusionary geopolitics through which they collectively set the terms of engagement between the 'Circumpolar' Arctic and the rest of the international community.

Global Arctic: Redux

The term 'Global Arctic' is gaining currency in contemporary debates about how the Arctic and the rest of the world are connected. Several academics have used it to foreground how what happens in the Arctic has global implications (in terms of its consequences for climate change, trade, commodity prices, or even the treatment of indigenous peoples, among other things), as well as how events in other parts of the world impact the Arctic (Dodds 2016; Shadian 2016; Heininen 2016). However, the term is not just an analytical tool for academics trying to study and explain the Arctic's relationship with the rest of the world. It has also gained a political currency which has subsequently been used to test the strength of circumpolarity as the defining ordering principle of Arctic affairs (Ingimundarson 2014). It has done so by exploiting shared topologies (the ways in which the Arctic and other parts of the world are scrumpled or folded together in networks of science, commerce, security, and conservation), rather than the topographies shared by the so-called Arctic states and in which 'circumpolar' claims are rooted.

When the A5 gathered first in Ilulissat in 2008, and then again in Ottawa in 2010, the most openly aggrieved state was Iceland. Iceland occupies a curious topographical position on the edge of the Arctic Circle. Its territory only just reaches above 66°N. However, its northern waters, as defined by its Exclusive Economic Zone (EEZ), are clearly part of the

Arctic. Nevertheless, because those waters fall short of the Arctic Ocean itself, Iceland is not considered a 'coastal' state by the rest of the A5—much to the chagrin of its Foreign Ministry, which has actively sought to construct an Arctic identity for the country based on it maritime interests (Dodds and Ingimundarson 2012). Iceland, which also had a deep economic crisis to contend with, responded to its exclusion by actively deepening its relations with non-Arctic states, especially China (Dodds 2015). Since 2004, Iceland's status as a transportation hub has been growing rapidly, with a huge increase in air traffic through the Keflavik international airport on the outskirts of Reykjavik (Gudjonsson and Nielsson 2015). However, its blossoming relationship with China has focused on science, resources, and shipping. In 2011, there was a joint Sino-Icelandic expedition to the North Pole; in 2012, the two countries signed a Framework Agreement on Arctic Cooperation, followed in 2013 by a Free Trade Agreement, the first between China and a European country. Crucially, Iceland also offers a 'stepping stone' for China to reach the mineral resources of Greenland (Degeorges 2014).

In April 2013 (on the same day that Iceland announced its Free Trade Agreement with China), Ólafur Ragnar Grímsson, the then-Icelandic President (1996–2016), took matters a step further by announcing that Iceland—in an attempt to cement a new position for the country as the meeting place of the world in the Arctic—would later that year begin hosting a new annual assembly of international Arctic stakeholders, provocatively named 'Arctic Circle'. He described the assembly's aim as being to 'strengthen the policymaking process [in/for the Arctic] by bringing together as many Arctic and international players as possible under one large tent' because 'the Arctic has suffered from a lack of *global* awareness and, as a result, a lack of effective governance' (Webb 2013, emph. added). Participants were to be drawn from across the world. The assembly would be a forum where virtually anyone who had something to say about the future of the Arctic could have a voice, from non-Arctic states to various foundations, companies, and NGOs which operate globally[15] (Depledge and Dodds 2017). Devoid of any topographical limits, Arctic Circle would be an early institutional manifestation of the idea of a 'Global Arctic'. Arctic Circle has met every year since, and has itself attempted to 'go global' by organising smaller, satellite forums in Greenland, Canada (Quebec), Singapore, and, most recently, Scotland. Annual attendance has been high, at more than 2000 participants from over 50 countries, including heads of states, government ministers, members of parliaments, officials, experts, scientists, entrepreneurs, business leaders, indigenous representatives, environmentalists, students, and activists.

The timing of Arctic Circle's launch in the run-up to the Kiruna Ministerial Meeting of the Arctic Council is unlikely to have been coincidental. The Arctic Council was due to decide on whether to allow China, Japan, Singapore, Japan, South Korea, India, and the EU, as well as several NGOs, to become Observers. As described elsewhere,

> there was palpable feeling at the time that Grímmson was challenging the Arctic Council to take on a more global profile, with a view that should the Arctic Council reject the observer applications, then the Arctic Circle would be prepared to provide an alternative platform for global interest to be expressed in the Arctic. (Depledge and Dodds 2017: 142–143)

Given that the Arctic Council had in fact spent several years deliberating on how to incorporate new Observers, it is unlikely that the launch of Arctic Circle was the deciding factor in the decision to admit six new states, but by the time of the Kiruna meeting, it was clear that two very different images of the Arctic—one rooted in circumpolarity, the other forged through global connectivity—were rubbing up against each other. What was perhaps most striking about Grímmson's initiative, then, was the provocation it posed to the other Arctic states, and especially the A5, as they attempted to maintain their primacy in Arctic affairs.

Ultimately, the Arctic Council's decision to accept the applications of six new non-Arctic states to become Observers to the Arctic Council made it harder to criticise the A8 for lacking awareness of, and openness to, growing international interest in the Arctic. Moreover, the format of Arctic Circle, which presented itself as a 'Davos for the Arctic',[16] suggested there was actually little intent by Grímsson and his partners to rival the Arctic Council, which is principally an intergovernmental forum supported by several scientific working groups. Instead, Arctic Circle, which has quickly become the largest Arctic-focused event in the world, has become a global meeting point for trading and exchanging ideas and information about the Arctic (Depledge and Dodds 2017).

Arctic Circle's impact on Arctic affairs has not yet been studied in-depth, but it is interesting to note how carefully non-Arctic states have engaged with it. Grímsson's partners in the venture included prominent representatives of Alaska, Greenland, and less explicitly the Faroe

Islands—a 'network of the marginalised' in the Arctic which seemed to be trying to develop new partnerships with the rest of the world in response to feeling that circumpolarity had failed them (ibid.). However, in Iceland's case, those anxieties appear to have been addressed for now by the rest of the Arctic Council's willingness to bring in Asian stakeholders. While Finland and Sweden were not closely involved with Arctic Circle, their joint interest in having the Arctic Council recognise the EU's interests in the Arctic was also progressed in Kiruna.[17] As a consensus-based organisation, the Arctic Council required that, following the Kiruna Ministerial, the A8 unite behind their joint decision. That move, together with admitting new Observer states to the Arctic Council, also reinforced the primacy of the Circumpolar Arctic states.[18] Aware that the Arctic Council, as opposed to Arctic Circle (or any other Arctic stakeholder forum), remains the primary forum with which to engage, non-Arctic states, including Britain, have been careful about what they say at Arctic Circle, typically couching their speeches in terms of their ongoing respect for the A8's sovereignty, sovereign rights, and jurisdictions in the Arctic, and the traditions and culture of Arctic indigenous peoples.[19]

Following the divisions created by the Ilulissat Declaration, the Arctic Council was arguably strengthened as an institution, as the Arctic states have once again found common 'circumpolar' cause on a range of issues from search and rescue, to oil spill response, to scientific cooperation. The A5 still convenes, albeit in a different form, its focus now on showing leadership on fisheries management in the high seas area of the Central Arctic Ocean where there is no sovereignty impediment to cooperating with Iceland and other major fishing nations such as China or the EU.

However, the annual meetings of Arctic Circle and its satellite forums continue to act as a persistent reminder of global interest in the Arctic and the threat it poses to circumpolar unity. So long as global interest in the Arctic persists, Arctic states that feel that their interests are no longer being achieved through circumpolar cooperation may well be tempted to reach beyond the region in search of new partnerships forged through the construction of shared topologies (scientific, commercial, etc.), rather than topographies. That already seems evident from the rapid increase in commercial ties between Russia and China in the Arctic after the other Arctic states joined the United States/EU in imposing sanctions on economic activity in the Russian Arctic in response to Russia's annexation of Crimea in 2014.

Britain: Pushed to the Periphery

The circumpolarisation of the Arctic in the twentieth century produced several challenges to British engagement with the region. Maps of the Arctic invariably pushed Britain to the periphery. If Britain appeared at all, it was at the extremity. When seen as part of a longer cartographic/carto-political tradition to position one's own country at the centre of the map to illustrate power and influence, that has the performative effect of distancing Britain topographically from the centre of the Arctic and the international waters of the Central Arctic Ocean, while enabling the A8 to question whether Britain has any right to claim a voice in Arctic affairs at all.

Setting aside such 'cartopolitics',[20] the institutionalisation of circumpolarity in the form of the Arctic Council and subsequent efforts by the A8 to position it as the primary intergovernmental forum for dealing with Arctic affairs also posed a more practical challenge to British engagement with the region: the need to respect the primacy of the Arctic states in circumpolar affairs as a precondition for being able to participate in the work of the Arctic Council.

In the 1990s and early 2000s, successive British governments appeared to be largely ambivalent about being on the periphery of Circumpolar Arctic institutions such as the Arctic Council. The post-Cold War peace dividend had brought security to the Northern Flank, and although Britain still had extensive scientific, environmental, and, to a lesser extent, commercial interests in the Arctic, it was unclear how those connections could be pursued.

Britain faced two constraints. The first was that there was apparently no strategy in the British Government for how Britain should pursue its interests in the Arctic, or even a clear understanding of what those interests were. Until 1990 the Arctic had been covered by the Maritime and Aviation Department. The brief was then passed to the Environment, Science and Energy Department formed in the run-up to the 1992 UN Conference on Environment and Development. Meanwhile, British representation in the IASC negotiations was managed by the National Arctic Research Forum (NARF), a separate joint Foreign Office/Natural Environment Research Council (NERC)-led cross-government initiative with responsibility for promoting British Arctic research, briefing delegates ahead of international meetings, and liaising between different departments and agencies with Arctic research interests.

By the mid-1990s, the NARF was effectively dissolved, depriving the British Government of any mechanism for coordinating cross-departmental interest and activity in the Arctic. The Environment, Science and Energy Department was preoccupied with international negotiations relating to the emerging conventions on biodiversity and climate change. The absence of ministerial interest or even top-down direction from John Major's Government (1992–1996) was evident in the bottom-up approach that Dr. Mike Richardson, the head of the Polar Regions Section in the Foreign Office, which was traditionally occupied with administering Britain's Antarctic policies, felt compelled to take in 1995, when he brought the Arctic within his team's remit.

Richardson, it seems, had feared that without a new coordinating mechanism to replace the NARF, Britain would find it hard to capitalise on the limited human and financial resources that the Government had made available for Arctic affairs, potentially costing Britain the Observer status it had worked hard for in AEPS. In the absence of high-level support, Richardson decided that the Polar Regions Section had to take a bottom-up approach to assessing the priorities and interests of different departments if a more coherent view of Britain's Arctic interests was to be produced. That was necessary for ensuring that, even in the absence of top-level interest in the Arctic, Britain maintained a strong presence in the Arctic, despite slender human and financial resources. It was clear, to Richardson at least, that Britain still retained scientific, commercial, and strategic interests in the Arctic. He also believed that it was still possible that, supported by the United States, Britain could secure an elevated role in Arctic affairs relative to other non-Arctic states (Northern Waters Group 1996).

The other source of constraint on Britain was external. Wider commentary from the time emphasise the 'growing pains' that afflicted AEPS and the Arctic Council in the 1990s. Britain was active in these negotiations to the extent that it sought to protect, at the very minimum, its Observer status. However, in a detailed report assessing the organisation behind AEPS, Håken Nilson of the Norwegian Polar Institute found that the principal problem facing Britain and other non-Arctic states was that they were seeing too little return on the scientific resources they were investing into the AEPS working groups. Participation was costly, as it typically required sending scientists and other representatives to the Arctic regions where meetings were held. Britain, for instance, had only been able to attend some of the AEPS meetings from the early to mid-1990s. Even when they did attend, their participation at the policy level was heavily curtailed. For example, it was considered inappropriate for Britain and other non-Arctic states to send Ministers. At the same time, the Observer states themselves could not agree on what their role should be, precluding any attempt to improve their standing collectively (Nilson 1997).

David Scrivener, a British scholar working on Arctic issues in the 1980s and 1990s, observed that even in the late 1990s, the A8 were still bogged down in Arctic Council negotiations over matters of procedure. That in turn curtailed the activities of the working groups (Scrivener 1999). Furthermore, he noted that: 'almost two years into its existence, the Arctic Council has so far very little to show for itself… [it was] still in its "peri-natal" phase' (ibid.: 57). Oran Young, the American doyen of Arctic governance studies, echoed Scrivener when he wrote that it might be 'best to treat [the Arctic Council] as a proto-regime that may or may not evolve over time into a fully-fledged regime for the circumpolar world' (Young 2000: 6). Richardson's view seems to have been that, despite these 'growing pains', it was imperative that Britain was well positioned to pursue its interests if and when those activities progressed into something more concrete.

Another issue in the 1990s was that much of the early work of the Arctic Council focused on local and transnational environmental problems and the welfare of indigenous peoples—issues where Britain, owing to its industrial and imperial legacies, was more likely to be seen as part of the problem (for example, as a major source of radioactive contamination from sites such as the Sellafield nuclear fuel reprocessing plant).[21] Whether Richardson anticipated a future surge of interest in the Arctic along the lines witnessed over the past decade or so is unclear, although even if he did, there were only limited attempts by successive British governments in the 1990s to push back against the 'circumpolarisation' of the Arctic. Perhaps that was because from the pre-2000 vantage point of Richardson and others, significant intergovernmental cooperation in the Arctic, whether on circumpolar terms or not, still seemed an unlikely prospect, especially given the United States' intransigence over allowing the Arctic Council to become anything more than a talking shop. Had they known that the Arctic Council would evolve rapidly over the next decade to become the principal arbiter of the relationship between Arctic and non-Arctic states, and, at the same time, that commercial interest in the Arctic would be resurgent alongside growing anxieties about climate change, Richardson and others might have had a stronger case to put before Government, about the need to properly fund British engagement with the Arctic Council and the region in general.

An opportunity to strengthen Britain's position within the Arctic Council came in the early 2000s during the planning phases of ACIA. ACIA began as a response to a recommendation by the IASC that the Arctic Council undertake an assessment of climate change in the Arctic to inform the work of the UN IPCC. The United States, which was the chair of the Arctic Council at the time, volunteered to lead the effort and, under its direction, the Arctic Council agreed to open up the process so that non-Arctic states with considerable expertise in Arctic research could be invited to make a substantial contribution. At the first scoping workshop, Britain and Germany were singled out as two countries in particular that had experts who could contribute to the study (ACIA 2000). By the end of the second meeting, two British scientists had been invited to serve as lead authors (the only lead authors from non-Arctic states). One of those, Terry Callaghan, was to be funded by Sweden where he worked. The other, Mark Nuttall, received support from the Polar Regions Section. According to a Foreign Office official who was present at the time, the decision to fund the participation of a British scientist as lead author was in part to show that the Arctic still mattered to Britain, particularly as part of broader British Government objectives relating to global climate change, and also to demonstrate Britain's desire to engage with the work of the Arctic Council.[22]

The impact of Britain's contribution to ACIA should not be exaggerated. It represented a modest commitment from the British Government to support the work of the Arctic Council, and also signalled that the A8 recognised the value of British polar science. But ACIA was ultimately owned by the Arctic Council and A8. Moreover, as ACIA sparked a step change in the level of worldwide interest in the Arctic, the gains that Britain might have made at the Arctic Council were minor in comparison to the impact of overwhelming international interest in the Arctic. That interest has since pushed the A8 to consolidate their shared circumpolarity within the Arctic Council's institutional framework and rules of procedure, leaving Britain and others to rethink how their relationship with the Arctic Council should evolve in the future.

Conclusion

By the end of the twentieth century, 'circumpolarity' had emerged as a powerful ordering principle in Arctic geopolitics—one which continues to define Arctic affairs today. The strength of circumpolarity is rooted in the emergence of the Arctic Council as the primary intergovernmental forum for Arctic affairs. Although in practice the Arctic Council is limited by its tight mandate and consensus-based approach, it has become an important body for at least two reasons: the value of its scientific assessments to global efforts to address challenges such as climate change and environmental pollution; and its emergence as a vehicle for the Arctic states to address common challenges relating to growing human activity within their respective jurisdictions. These two developments, combined, perhaps, with the arresting name of the organisation ('Arctic Council'), have driven non-Arctic states, including Britain, seeking to play an active role in the region, to show the A8 a high degree of deference in return for being able to attend meetings and take part in the scientific working groups.

However, as negotiations about the terms of reference and rules of procedure that would inform first AEPS and later the Arctic Council dragged on, countries such as Britain could also be forgiven for their apparent ambivalence about attempts to establish a circumpolar order, not least because the Arctic states themselves did not seem to agree on whether such an order was desirable. The United States, in particular, pushed hard for countries such as Britain and the Netherlands to be recognised as important Arctic actors. The Arctic was not a priority for much of the British Government and although NERC, the Foreign Office, and other parts of the NARF clearly wanted to maintain a scientific presence in the Arctic and saw value in being part of the new institutional landscape for international scientific cooperation (rooted in IASC, AEPS, and eventually the Arctic Council), they were not prepared to do so at any cost. In the end, Observer status appears to have been accepted as a way of keeping a 'strong foot in the door', rather than as a mechanism for pushing British interests (which, aside from science, were relatively diminished in the 1990s) or challenging the emerging circumpolar order.

In Britain's case, where successive governments since the early 1990s have been largely ambivalent about Arctic issues, and engagement has mostly relied on bottom-up leadership from the Polar Regions Section (now the Polar Regions Department) with limited human and financial resources, that seems to have been a price worth paying. At the very least, it has ensured Britain has retained access to circumpolar institutions, although Foreign Office officials will likely also argue that by supporting circumpolar cooperation, Britain has helped to maintain peace and stability in a region which has long been of strategic concern, without having to invest too many of its own resources.

Yet in the first decade of the twenty-first century, the attitude of the eight Arctic states appears to have hardened. The success of ACIA demonstrated the Arctic Council's value as a mechanism for circumpolar cooperation even though several non-Arctic states, including Britain, were also involved. Meanwhile growing interest in and speculation about the Arctic worldwide—much of which was wildly out of line with reality—gave those with a more exclusive view of circumpolar affairs reason to seek a clearer divide between the 'Arctic states' and the 'rest of the world', defined, topographically, by a country's position in relation to the Arctic Circle.

The problem for Britain and other non-Arctic states with strong interests in the Arctic is that this topographically rooted circumpolarity does not easily accommodate claims of topological proximity that link the fates of security, commerce, and environmental sustainability in otherwise distant places. Relative topographical proximity to the region matters far less, so while Britain and China, for example, may claim some kind of 'near-Arctic' status, as a way of seeking privileged access to the region vis-à-vis other non-Arctic states, in practice it does not bestow on them any more right to be part of Arctic affairs than any other non-Arctic state. To the A8, the topography of everything below the Arctic Circle is largely irrelevant. Perhaps if the topographical case had been made sooner, during the negotiations of the IASC or AEPS and the Arctic Council, it would have been easier to stretch what constitutes the Arctic to encompass 'near-Arctic' interests. Instead, this language appears to be relatively novel—and even slightly desperate—as non-Arctic states jostle to position themselves as a premier partner for Arctic states and other regional stakeholders, although, as yet, to little obvious effect.

Whether 'circumpolarity' will remain the defining ordering principle of Arctic geopolitics is still to be seen. Today, the process of 'circumpolarisation' remains incomplete, in part because the nineteenth-century imaginary of a 'Wider', and increasingly 'Global' Arctic has never gone away. As with all geographical constructs, there is an inherent 'stretchability' about the Arctic which means that its borders retain the potential to continue shifting over time, especially as new topologies connecting the Arctic with other parts of the world are forged, whether for environmental, technological, scientific, economic, strategic, or demographic reasons. For Britain and other non-Arctic states seeking a greater presence in the region, it is not through recourse to their topography, but through putting greater emphasis on the topologies that they share—or which might be forged—with the Arctic (past, present, and future) that they are most likely to legitimise and promote their interests. Only by showing how they continue to be folded together with the Arctic—through connections brought about by science, defence, commerce, pollution, and so on—will countries such as Britain be able to demand and justify greater involvement in Arctic affairs.

Notes

1. Both connection and disconnection, topologically speaking, should therefore be regarded as constituted by socio-material processes and not geographical conditions.
2. A third IPY was held in 1957–1958, although its broader scientific mandate led to it being renamed the International Geophysical Year (or IGY).
3. It is worth noting that among indigenous peoples of the Arctic, there was also great interest in circumpolarity—many of these groups had arbitrarily been separated from one another by the imposition of geopolitical borders in the Arctic. The Inuit Circumpolar Council established in 1977 was responsive to this.
4. Ice, land, and water have all been treated differently in international law. These themes have been explored recently by the ICE LAW Project led by Durham University (www.icelawproject.org).
5. Another term which has proven 'stretchable' over the years as countries such as Britain and China have sought to highlight their proximity to the Arctic.
6. Initially, there was to be a non-government scientific organisation and a separate intergovernmental forum but the latter was dropped, resulting in governments taking greater interest in the development of IASC.
7. Informally, Arctic affairs are also shaped by other sites of knowledge exchange such as international policy conferences (see Depledge and Dodds 2017).
8. Both the Central Arctic Ocean and the deep sea bed beyond continental shelves fall outside the national jurisdictions of the A8 but the possibility of commercial exploitation in these areas is likely to still be hampered by hazardous environmental conditions for some time yet.
9. See Chap. 4.
10. For a fascinating account of the expedition, see McDowell and Batson (2007).
11. The decision by Russia to restart the Soviet-era practice of sending its bomber aircraft on long-range flights, including over the North Pole only added to the hype that started to build around the Arctic in 2007 (BBC News 2007b).

12. Scott Borgerson description of 'the coming anarchy' in the Arctic may well have been nodding to Robert Kaplan's (1994) famous essay of the same title published in *The Atlantic* in 1994, which had painted a dire vision for the world's future in which resource scarcity and climatic upheaval had 'destroyed the social fabric of our planet' (Kaplan 1994).
13. While the European Parliament was not alone in calling for a new legally binding instrument to govern Arctic affairs it attracted the most attention, not least because tensions between the EU and Canada were already in the spotlight over the seal ban issue.
14. Author interview with Foreign Office official, June 20, 2013.
15. Examples include the Mamont Foundation, Guggenheim, MacArthur Foundation, WWF, Prince Albert II of Monaco Foundation, European Climate Foundation, and the Carnegie Corporation of New York.
16. Davos being a reference to the annual meeting of the World Economic Forum in Davos, Switzerland, which brings together top business leaders, political leaders, intellectuals, and journalists to discuss pressing global issues.
17. The EU's application was received 'affirmatively' at the Kiruna Ministerial but a final decision was not to be implemented until the seal ban issue with Canada was resolved.
18. Not being states, Alaska, Greenland, and the Faroe Islands were ultimately subject to the decisions of their national governments.
19. The British delegation likely only got away with a more exuberant performance in 2014 because it did not officially represent the Government. The Foreign Secretary Philip Hammond's video message at the same event was far more cautious. For an account see Depledge (2014). No country, including Britain, has attempted anything similar since.
20. For more on Arctic 'carto-politics' see the work of the Danish political scientist Jeppe Strandsbjerg (2012).
21. In 2002, the Arctic Council's Arctic Monitoring and Assessment Programme (AMAP) recommended that the Arctic Council encourage Britain to reduce releases from Sellafield. The Arctic Council's subsequent 'Inari Declaration' in 2002 was a little softer, urging all non-Arctic European states to take action to reduce releases of radioactivity from reprocessing facilities.
22. Author interview with Foreign Office official, 19 October 2011.

References

ABA. 2013. Arctic Biodiversity Assessment 2013. *Arctic Council.* Accessed June 22, 2017. https://oaarchive.arctic-council.org/handle/11374/133

ACIA. 2000. Arctic Climate Impact Assessment: Report on the 3rd Meeting of the Assessment Steering Committee and a Scoping Workshop. *ACIA.* Accessed June 22, 2017. http://www.acia.uaf.ed/PDFs/Scoping-Report.pdf

———. 2004. *Impacts of a Warming Arctic: Arctic Climate Impact Assessment.* Cambridge: Cambridge University Press.

AMSA. 2009. Arctic Marine Shipping Assessment Report. *Arctic Council.* Accessed June 22, 2017. https://oaarchive.arctic-council.org/handle/11374/54

Arctic Council. 2013. Arctic Council Observer Manual for Subsidiary Bodies. *Arctic Council.* Accessed June 22, 2017. https://oaarchive.arctic-council.org/handle/11374/939

ARR. 2016. Arctic Resilience Report. *Arctic Council.* Accessed June 22, 2017. https://oaarchive.arctic-council.org/handle/11374/1838

BBC News. 2007a. Russians to Dive Below North Pole. *BBC News*, July 24. Accessed January 6, 2017. http://news.bbc.co.uk/1/hi/world/europe/6914178.stm

———. 2007b. Russia Restarts Cold War Patrols. *BBC News*, August 17. Accessed January 6, 2017. http://news.bbc.co.uk/1/hi/world/europe/6950986.stm

Borgerson, Scott. 2007. An Ice Cold War. *The New York Times*, August 8. Accessed January6,2017.http://www.nytimes.com/2007/08/08/opinion/08borgerson.html

———. 2008. Arctic Meltdown. *Foreign Affairs* March/April. Accessed October 31, 2016. https://www.foreignaffairs.com/articles/arctic-antarctic/2008-03-02/arctic-meltdown

Carter, Lee. 2007. Arctic Neighbours Draw Up Battle Lines. *BBC News*, August 11. Accessed January 6, 2017. http://news.bbc.co.uk/1/hi/world/americas/6941569.stm

CBC. 2009. Canada Against EU Entry to Arctic Council Because of Seal Trade Ban. *CBC News*, April 29. Accessed January 5, 2017. http://www.cbc.ca/news/canada/north/canada-against-eu-entry-to-arctic-council-because-of-seal-trade-ban-1.806188

Chang, Gordon. 2010. China's Arctic Play. *The Diplomat*, March 9. Accessed July 6, 2017. http://thediplomat.com/2010/03/chinas-arctic-play/

Degeorges, Damien. 2014. Greenland and Iceland: Meeting Place of Global Powers in the Arctic. *IFRI*, September 4. Accessed January 6 2017. https://www.ifri.org/en/publications/editoriaux/actuelles-de-lifri/greenland-and-iceland-meeting-place-global-powers-arctic

Depledge, Duncan. 2014. You're Only Supposed to Blow the Bloody Doors Off. *RHUL Geopolitics & Security*. Accessed July 6, 2017. https://rhulgeopolitics.wordpress.com/2014/11/03/youre-only-supposed-to-blow-the-bloody-doors-off/

Depledge, Duncan, and Klaus Dodds. 2017. Bazaar Governance: Situating the Arctic Council. In *Governing Arctic Change*, ed. Kathrin Keil and Sebastian Knecht, 141–160. Basingstoke: Palgrave Macmillan.

Dodds, Klaus. 2002. *Pink Ice: Britain and the South Atlantic Empire.* London: I.B. Tauris.

———. 2008. Icy Geopolitics. *Environment and Planning D: Society and Space* 26: 1–6.

———. 2015. From Ilulissat to Kiruna: Managing the Arctic Council and the Contemporary Geopolitics of the Arctic. In *Handbook of the Politics of the Arctic*, ed. Leif Christian Jensen and Geir Hønneland, 375–387. Cheltenham: Edward Elgar.

———. 2016. What We Mean When We Talk About the Global Arctic. *Arctic Deeply*, February 18. Accessed January 6, 2017. https://www.newsdeeply.com/arctic/community/2016/02/18/what-we-mean-when-we-talk-about-the-global-arctic

Dodds, Klaus, and Valur Ingimundarson. 2012. Territorial Nationalism and Arctic Geopolitics: Iceland as an Arctic Coastal State. *The Polar Journal* 2: 21–37.

European Commission. 2008. *Climate Change and International Security*. European Commission. Accessed June 22, 2017. https://www.consilium.europa.eu/uedocs/cms_data/docs/pressdata/en/reports/99387.pdf

European Parliament. 2008. *European Parliament Resolution of 9 October 2008 on Arctic governance*. European Parliament. Accessed January 5, 2017. http://www.europarl.europa.eu/sides/getDoc.do?type=TA&reference=P6-TA-2008-0474&language=EN

Foxall, Andrew. 2014. 'We Have Proved It, the Arctic is Ours': Resources, Security and Strategy in the Russian Arctic. In *Polar Geopolitics? Knowledges, Resources and Legal Regimes*, ed. Richard Powell and Klaus Dodds, 93–112. Cheltenham: Edward Elgar.

Gorbachev, Mikhail. 1987. Mikhail Gorbachev's Speech in Murmansk at the Ceremonial Meeting on the Occasion of the Presentation of the Order of Lenin and the Gold Star to the city of Murmansk. *Barents Info*, October 1. Accessed December 22, 2016. https://www.barentsinfo.fi/docs/Gorbachev_speech.pdf

Graczyk, Piotr, and Timo Koivurova. 2014. A New Era in the Arctic Council's External Relations? Broader Consequences of the Nuuk Observer Rules for Arctic Governance. *Polar Record* 254: 225–236.

Gudjonsson, Heidar, and Egill Thor Nielsson. 2015. Iceland's Arctic Awakening. *Arctic in Context*, April 22. Accessed January 6, 2017. http://www.worldpolicy.org/blog/2015/04/22/iceland%E2%80%99s-arctic-awakening

Heininen, Lassi. 2016. High Arctic Stability as an Asset for Storms of International Politics – An Introduction. In *Future Security of the Global Arctic: State Policy, Economic Security and Climate*, ed. Lassi Heininen, 1–11. Basingstoke: Palgrave Macmillan.

Ilulissat Declaration. 2008. *Arctic Governance Project*. Accessed December 30, 2016. http://www.arcticgovernance.org/the-illulissat-declaration.4872424.html

Ingimundarson, Valur. 2014. Managing a Contested Region: The Arctic Council and the Politics of Arctic Governance. *The Polar Journal* 4: 183–198.

Kaplan, Robert. 1994. The Coming Anarchy. *The Atlantic*, February. Accessed January 6, 2017. http://www.theatlantic.com/magazine/archive/1994/02/the-coming-anarchy/304670/

Kristoffersen, Berit. 2014. 'Securing' Geography: Framings, Logics and Strategies in the Norwegian High North. In *Polar Geopolitics? Knowledges, Resources and Legal Regimes*, ed. Richard Powell and Klaus Dodds, 131–148. Cheltenham: Edward Elgar.

Lucas, Edward. 2008. *The New Cold War: How the Kremlin Menaces Both Russia and the West*. London: Bloomsbury Publishing.

McDowell, Mike, and Peter Batson. 2007. *Last of the Firsts: Diving to the Real North Pole*. The Explorer's Club. Accessed June 22, 2017. https://explorers.org/flag_reports/Mike_McDowell_Flag_42_Report.pdf

Nilson, Håken. 1997. *Arctic Environmental Protection Strategy (AEPS): Process and Organization, 1991–1997. Norwegian Polar Institute*. Accessed June 22, 2017. https://brage.bibsys.no/xmlui/handle/11250/173498

Northern Waters Group. 1996. *Northern Waters and Arctic Study Group Meeting*. Minutes of the Northern Waters and Arctic Study Group Meeting, Chatham House, April 22.

Oslo Agreement. 1973. Agreement on the Conservation of Polar Bears. *Norwegian Polar Institute*. Accessed December 22, 2016. http://pbsg.npolar.no/en/agreements/agreement1973.html

Ottawa Declaration. 1996. Declaration on the Establishment of the Arctic Council. *Arctic Council*. Accessed December 22, 2016. https://oaarchive.arctic-council.org/bitstream/handle/11374/85/EDOCS-1752-v2-ACMMCA00_Ottawa_1996_Founding_Declaration.PDF?sequence=5&isAllowed=y

Reuters. 2007. Canada Mocks Russia's "15th Century" Arctic Claim. *Reuters*, August 3. Accessed January 6, 2017. http://uk.reuters.com/article/uk-russia-arctic-canada-idUKN0246498520070802

Scrivener, David. 1999. Arctic Environmental Cooperation in Transition. *Polar Record* (192): 51–58.

Shadian, Jessica. 2016. Finding the Global Arctic. *The Arctic Journal*, March 14. Accessed January 6, 2017. http://arcticjournal.com/opinion/2216/finding-global-arctic

Smieszek, Malgorzata. 2015. 25 Years of the International Arctic Science Committee (IASC). In *Arctic Yearbook 2015*, ed. Lassi Heininen, Heather Exner-Pirot, and Joel Plouffe. Akureyri: Northern Research Forum. Accessed January 17, 2017. http://www.arcticyearbook.com

Steinberg, Philip, Jeremy Tasch, Hannes Gerhardt, Adam Keul, and Elizabeth A. Nyman. 2015. *Contesting the Arctic: Politics and Imaginaries in the Circumpolar North*. London: I.B. Tauris.

Stepień, Adam, and Timo Koivurova. 2015. The Making of a Coherent Arctic Policy for the European Union: Anxieties, Contradictions and Possible Future

Pathways. In *The Changing Arctic and the European Union*, ed. Adam Stepień, Timo Koivurova, and Paula Kankaanpää, 20–56. Leiden: Brill Nijhoff.

Stokke, Olav Schram. 1990. The Northern Environment: Is Cooperation Coming? *The Annals of the American Academy of Political and Social Science* 512: 58–68.

———. 2011. Environmental Security in the Arctic: The Case for Multilevel Governance. *International Journal* 66: 835–848.

Strandsbjerg, Jeppe. 2012. Cartopolitics, Geopolitics and Boundaries in the Arctic. *Geopolitics* 17: 818–842.

Webb, Robert. 2013. Iceland President Sounds Climate Alarm Demanding Global Attention, Action at NPC Luncheon. *The National Press Club*, April 15. Accessed January 6, 2017. https://www.press.org/news-multimedia/news/iceland-president-sounds-climate-alarm-demanding-global-attention-and-action

Woods, Allan. 2010. Canada Gets Cold Shoulder at Arctic Meeting. *Toronto Star*, March 29. Accessed December 30, 2016. https://www.thestar.com/news/canada/2010/03/29/canada_gets_cold_shoulder_at_arctic_meeting.html

Young, Oran. 2000. *The Structure of Arctic Cooperation: Solving Problems/Seizing Opportunities*. Paper presented at the Fourth Conference of Parliamentarians of the Arctic Region, Rovaniemi, August 27–29.

———. 2005. Governing the Arctic: From Cold War Theater to Mosaic of Cooperation. *Global Governance* 11: 9–15.

CHAPTER 4

Britain in the Arctic Today

Abstract For nearly two decades after the end of the Cold War, successive British governments appeared largely ambivalent about Britain's interests in the Arctic. That changed in the mid-2000s, as concerns about climate change, energy security, and Russia, brought the Arctic in from the cold. Since then, British interest has grown in at least three areas: science, defence, and commerce. This chapter explores where these interests come from, what actors are involved, and how they are forging new connections between Britain and the Arctic.

Keywords Britain and the Arctic • Science • Defence • Commerce • Climate change • Contemporary Arctic

In the decade since Russia's flag was planted at the North Pole, British interest in the Arctic has returned to a level not seen since the Cold War. Successive governments from the early 1990s to the mid-2000s were largely ambivalent about British engagement with the Arctic. But that period of diminished interest also tells us something important: British interest in the Arctic is not self-evident just because, topographically, Britain is nearer to the Arctic than any other non-Arctic country. Rather, British Arctic interest remains fluid, shaped by specific interests emerging at specific times. Both history and geography are therefore precarious means by which to determine Britain's contemporary Arctic interests.

© The Author(s) 2018

D. Depledge, *Britain and the Arctic*,
https://doi.org/10.1007/978-3-319-69293-7_4

Another danger is that rooting contemporary British Arctic interests in a sense of historical or geographical permanency risks the criticism that Britain continues to see the Arctic in colonial or neocolonial terms: as a region where it is entitled to project its influence on the basis of its past history and topographical proximity. Polar law scholars Rachael Johnstone and Federica Scarpa (2016: 7) have described such claims—which resonate in other European capitals as well—as a kind of 'State-level snobbery of "old money" versus "new money"' intended to set European states apart from 'Asian interlopers' (which may lack a deep history in the Arctic or are more distant topographically). They go on to ask 'why historical connections are still seen as so important' when the real 'question for Arctic relations today is what a state (or other non-Arctic actor) can offer *now*, not what it did a century ago' (ibid.: 7). And they remind European governments in particular that in addition to their citizens' heroic feats in the Arctic, it is their industries which must carry a large proportion of the blame for environmental pollution, destruction of ecosystems, and climate change in the Arctic—issues on which those governments seemingly prefer to be future-facing.

Over the past decade it has been possible to discern, from the policy statements, speeches, and actions considered below, that successive British governments have had at least three broad national priorities in the Arctic: (1) to monitor geopolitical developments that might be relevant to British defence and security planning; (2) to enhance the British scientific community's contribution to, and address key uncertainties in, Arctic and climate change science; and (3) to facilitate British-based industry in pursuing new commercial opportunities in the region. As successive governments have acted to address those priorities, partly in tandem with a variety of other British-based Arctic stakeholders, new topologies have been forged with the Arctic that, collectively, have produced a renewed sense of proximity to the Arctic that is only partially rooted in, and certainly not determined by, topography.

The focus here, then, is on what the contemporary British interest in defence, science, and commerce in the Arctic is, where that interest comes from, and what actors are involved, as well as the topologies (i.e. connections between Britain and the Arctic) that are being forged in the process.

Maintaining Britain's Security

As far back as the Napoleonic Wars (1799–1815), British forces have conducted military operations in Arctic waters off the coast of Norway and Russia. During the Crimean War (1853–1856), the Royal Navy bombarded

Russian bases in the Arctic (Mazo 2015). During the First and Second World Wars, Britain battled to protect its sea lines of communication with North America, Russia, and later the Soviet Union from German interdiction. After the Second World War, the creation of the North Atlantic Treaty Organization (NATO) in 1949 entrenched the United States firmly in the Western European security order, and reaffirmed the strategic importance of the 'transatlantic bridge' via which the United States would reinforce its allies in Europe in the case of a military crisis with the Soviet Union.

That bridge was subsequently threatened throughout the rest of the Cold War by the Soviet submarine forces based in the Arctic. Britain and the United States responded by installing listening arrays across the North Atlantic, and embarking on 'under-the-ice' submarine operations which by the mid-1980s had pushed the 'Arctic' front line with the Soviet Union back to the Barents Sea. Meanwhile, British land forces maintained a continuous winter-time presence in Norway to deter a Soviet land invasion. Britain's stake in that strategy, which lasted until the Cold War ended, was specifically rooted in concerns about Britain's proximity to the Arctic theatre, and the possibility of using military forces to push that theatre's borders further away. As Admiral Sir William Staveley (First Sea Lord and Chief of the Naval Staff, 1985–1989) spelled out at a security conference in London in 1986:

> Consider the situation if we were to relax our guard in this strategically important area, putting at risk the sparsely populated region of North Norway, then Iceland and the Faeroes and thus placing the North Sea and the United Kingdom so much closer to the front line of Soviet forces, needlessly exposing ourselves to a greater threat. (Staveley 1988: 68)

Observing these developments, the Northern Waters and Arctic Study Group at the Scottish Branch of the Royal Institute of International Affairs identified that Britain's principal strategic interest in the Arctic were the warmer waters of the North Atlantic and European Arctic that bordered the Arctic Ocean (Archer and Scrivener 1986). However, these waters also mattered for economic reasons, not least because of the potential risk that disputes over resources posed to NATO unity, as demonstrated by the so-called 'Cold Wars' between Britain and Iceland in the 1950s and 1970s (Jónsson 1982).

After the Cold War ended, the Arctic frontline between NATO and the Soviet Union in the collapsed. Russia subsequently reduced its military presence along much of its Arctic coastline. Its remaining forces were concentrated on the Kola Peninsula to protect its nuclear weapons bases and remnants of the Northern Fleet. Britain and the United States responded by curtailing their own Arctic military footprints as part of a wider NATO drawdown of forces. As the West encouraged Russia to become part of a post-Cold War rules-based order, circumpolar cooperation and institution-building in the Arctic increased around common interests relating to science, the environment, and sustainable economic development. The peace dividend even brought about cooperative military initiatives, such as the Arctic Military Environmental Cooperation (AMEC) in 1996 between the United States, Norway, and Russia, with Britain joining later, to decommission Russia's deteriorating nuclear submarine fleet. In 2000, the Royal Navy assisted the Russian Navy during efforts to rescue the crew of a stricken Russian submarine (the *Kursk*) after it sank in the Barents Sea.

After a brief period of optimism between 2003 and 2005, West–Russia relations deteriorated again. So-called 'Colour Revolutions' swept across former Soviet countries, NATO enlargement brought several Eastern and Central European countries within its security architecture, and Russian President Vladimir Putin lamented the geopolitical catastrophe that had been caused by the fall of the Soviet Union. At the same time, the impacts of climate change on the Arctic were becoming more apparent. But where much of the world saw the Arctic as a site of transboundary environmental vulnerability, Russia in particular was concerned that a new strategic vulnerability was also emerging, owing to the increased accessibility of Arctic waters adjacent to around a fifth of Russian territory. That territory was also home to vast stores of undeveloped resources, and as the Kremlin sought to wrest back control over natural resource management, extraction, and taxes, recognition there grew of the Arctic's importance to Russia's present and future economy (Foxall 2014).

Between 2004 and 2006, Russia restarted regular missile tests and strategic submarine patrols in the Arctic. In 2007, strategic bombers resumed long-range air patrols over the Arctic. In 2008, Russia announced it would further enhance its military activity in the Arctic through to 2020 to safeguard national interests, which included the protection of emerging shipping lanes, resources, and strategic bases along its entire Arctic coastline. Those plans have since resulted in the opening (or reopening) of numerous search and rescue stations, radar stations, airfields and frontier outposts, at least three brigade-sized deployments, large-scale military exercises, the commissioning of new and more capable submarines to replace its ageing fleet, and, more recently, the deployment of several precision-guided missile defence systems aimed at defending the Russian Arctic from air and sea incursions (Conley and Rohloff 2015).

Britain and the United States meanwhile were still reducing their military presence in the Arctic. In 2006, the United States withdrew from the Keflavik airbase in Iceland. Britain suspended under-the-ice submarine operations in 2007 following an explosion on board HMS *Tireless* during an under-ice exercise which killed two crew members. Around the same time, the more pressing needs of the wars in Afghanistan and Iraq led the Ministry of Defence to scale back the Royal Marines' annual Arctic warfare training exercises in Northern Norway.

There is still much debate about how to read these developments. On the one hand, there has been anxiety in all of the Arctic states, and in several non-Arctic states including Britain, about how Arctic environmental change is likely to affect existing national defence and security interests. Layered on top of A5 concerns about how to enforce sovereignty as Arctic waters becomes more accessible are broader concerns, relevant to the A8 and others, about how to manage increases in commercial activity, tourism, and various illicit activities ranging from protests, to smuggling, illegal fishing, dumping of waste, and potentially even terrorism. In response, the A8 are all taking steps to upgrade and enhance their military and constabulary forces in the Arctic (Mazo and Le Miere 2013).

On the other hand, as a classic security dilemma has emerged it is difficult to determine what military activity in the Arctic is directed towards the maintenance of national securityA8 and what is directed towards wider geopolitical interests (Åtland 2014). For instance, the West's anxieties about Russian military activity in the Arctic are closely associated with a chain of events across Eastern Europe over the past decade—including Russia's 'energy wars' with Belarus and Ukraine between 2004 and 2010, its massive cyber-attack against Estonia in 2007, its war with Georgia in 2008, and its annexation of Crimea in 2014 and, latterly, its actions in the Middle East in support of the Syrian Assad regime—through which Russia has sought to pave the way for the emergence of a 'post-West' world order.[1] That has led some to ask whether the post-Cold War peace dividend in the North Atlantic and the High North is under threat (Olsen 2017).

Throughout the 1990s, Britain and the rest of NATO developed their defence and security policies under the implicit assumption that the trans-atlantic bridge no longer needed to be defended to the same extent that it was during the Cold War, primarily because the Arctic was no longer viewed as a source of military threat. British defence planners shifted their focus to developing expeditionary forces for discretionary military interventions around the world in places such as Iraq, Afghanistan, Bosnia, Kosovo, and Sierra Leone. Virtually the entire force posture of not just Britain, but also NATO, was shifted away from the defence and reinforcement of Europe towards so-called 'wars of choice' (Hudson and Roberts 2017).

The North Atlantic and Arctic were subsequently neglected by the military. The Royal Navy, devoid of its major role in maintaining the security of the North Atlantic and the High North, bore the brunt of military spending cuts. Specialist anti-submarine warfare platforms and maritime air patrol capabilities that were critically important during the Cold War were de-prioritised, while hunter-killer submarines were repurposed to provide sea-based power projection into areas of strategic interest. As the defence analysts Peter Hudson and Peter Roberts have argued: 'The perceived end of the Russian threat consigned the North Atlantic [and with it the Arctic] to a minor role in the eyes of the British military establishment' (Hudson and Roberts 2017: 82).

However, roughly a decade ago, despite the drawdown of military forces in the North Atlantic and High North, some forward-looking defence planners started looking at the Arctic again. Climate change emerged as a major policy issue for Britain under Tony Blair's Government, and several departments and agencies, including the Ministry of Defence and the Foreign Office, were asked to look closely at how climate change might impact Britain's interests overseas. The Arctic became a focal point for debates about the security implications of climate change, owing to warnings from defence experts about the potential for conflict over territory and resources as the region became increasingly ice-free. In 2007, the Development, Concepts and Doctrine Centre (DCDC), a Ministry of Defence think-tank based in Shrivenham, was tasked by the Ministry of Defence and the Foreign Office with examining these trends. While the content of DCDC's subsequent report remains restricted, its preparation signalled that at least some within the British Government were alert to the possibility that the Arctic was not necessarily as stable and secure as many would like to assume.

In the House of Commons too, the Shadow Defence Secretary Liam Fox and other Conservative MPs were also making their concerns about the Arctic known. In 2008, Fox warned that Russia was threatening to lay claim to a '460,000 square metre portion of ice-covered Arctic'. Fox called it a 'ridiculous' claim with no 'legitimate legal basis', but also warned that such threats still had to be taken seriously (Fox 2008). Another Conservative MP, Ian Taylor, also warned of an emerging Russian threat in the Arctic, telling the chamber that

> we forget that the Russians are also encroaching on the Arctic Circle, and any threat to Norwegian [energy] supply [to Britain] would be a grave problem for us. (Taylor 2008)

After the Russia–Georgia conflict that occurred just a few months later, speculation in the West about the potential for a conflict with Russia in the Arctic grew still further, prompting NATO and the Government of Iceland to jointly convene an Arctic security conference in Reykjavik. The British Defence Minister, Ann Taylor, told the meeting that Britain still had a significant stake in ensuring that the security and stability of the Arctic was upheld as new challenges and opportunities emerged in the region (Taylor 2009). But the resounding conclusion drawn from that conference was that the threat of conflict in the Arctic was being overstated (Holtsmark and Smith-Windsor 2009). Already embroiled in wars in Iraq and Afghanistan, it is scarcely surprising then that developments in the Arctic were doing little to alter the geopolitical mindset of British defence ministers towards the region.

However, a year later, the Ministry of Defence's attitude did start to change. Following the election of the Conservative–Liberal Democrat Coalition, Liam Fox, who had been arguing for several years that the defence community needed to take what was happening in the Arctic seriously, was appointed Secretary of State for Defence. Fox's impact was not immediate. The 2010 Strategic Defence and Security Review (SDSR) and the National Security Strategy did not include any references to the Arctic, while the military's most substantial contribution to defence in the North Atlantic and High North—its maritime patrol aircraft—was scrapped. Even so, Liam Fox did speak publicly about the neglect that Britain had shown in its own 'northern backyard' after years of expeditionary fighting in the Middle East. As Fox declared:

> We cannot forget that geographically the United Kingdom is a northern European country ... For too long Britain has looked in every direction except its own backyard. (Ministry of Defence 2010)

His language suggested that through a chronic lack of investment and attention, Britain, topographically that is, had effectively distanced itself from its strategic concerns in the Arctic and Northern Europe.

To reconnect Britain with the Arctic and Northern Europe, shortly after being appointed Defence Secretary, Fox visited Oslo to announce the launch of the Northern Group, a new forum for defence discussions consisting of Britain, the Nordic and Baltic countries, Germany, and Poland. *Contra* circumpolarity, the forum conjured up an image of a 'wider north' of strategic interest that wed strategic considerations in the Arctic to those in the Baltic, Northern Europe, and North Atlantic (Depledge and Rogers 2016). Britain was to have a major role in this theatre. The Ministry of Defence then signed a Memorandum of Understanding (MOU) with Norway to enhance bilateral defence cooperation in the Arctic. Defence officials at the time felt that a closer partnership with Norway was an obvious way for Britain to become more engaged in the Arctic. After all, Britain had a long-standing history of training for Arctic warfare in Norway, and, more recently, Britain had become increasingly reliant on Norway for energy. Resurgent concerns about Russia were also most likely to be felt in the European Arctic where Russia posed the greatest challenge.[2]

Fox's successors continued to reaffirm the importance of Britain's defence relationship with Norway. It was Philip Hammond (Defence Secretary, 2011–2014) who actually signed the MOU with Norway and continued to support the Northern Group. He was reportedly convinced of the importance of Britain's strategic interests in Norway and the High North by Sir Gerald Howarth, who served as a defence minister from 2010 to 2012 and provided an important source of continuity after Fox resigned. Sir Michael Fallon (Defence Secretary, 2014–present) has similarly continued to meet with Nordic and Baltic defence ministers for open and direct debate about defence cooperation in Northern Europe, including the Arctic. In 2016, Fallon, after visiting the Norwegian Arctic, signed a second MOU on defence cooperation with Norway to encourage closer cooperation on maritime patrol aircraft and exercises. Although Britain's maritime patrol aircraft were scrapped in the 2010 SDSR, the 2015 SDSR announced that the capability was to be restored through the procurement of nine Boeing-P8s, the same airframe that Norway is buying. Both countries will most likely use this capability to monitor Russian naval activity in the North Atlantic and High North. In 2017, Britain, alongside the United States, also accepted an invitation from Norway to take part in its national Exercise JOINT VIKING for the first time, with the intent of increasing cooperation between the three countries' armed forces. That exercise takes place annually in Norway's northernmost county and involves drills to exercise crisis management and the defence of Norway.

Ask at the Foreign Office or the Ministry of Defence today, and officials there will confirm that the British Government's highest priority in the Arctic is to support continued peace and stability. But that claim means little on its own. After all, continued peace and stability in the Arctic is in everyone's interests, not least because it underpins a broad array of other interests relating, for example, to science and commerce. Instead, what the Government should be saying is that Britain's stake in Arctic peace and stability is rooted in concerns about the elasticity of the Arctic's boundaries; that when it comes to security and defence, Britain's topographical location and the strength of its armed forces demands its inclusion in the strategic picture. In other words, if one was seeking an overview of the strategic picture of the Arctic in polar projection (i.e. as if looking at a map of the world with the North Pole at its centre), Britain's proximity—rooted in both its topographical position and its topological capability to project force into the Arctic—makes it difficult to exclude.[3]

However, the British Government's interest in continued peace and stability in the Arctic is also being communicated with restraint. Domestically, there is of course a risk that overemphasising Britain's defence interests in the Arctic could end up highlighting Britain's weaknesses with regard to enforcing peace and stability in the region. While Britain still has the most powerful military in Europe, those elements of its forces that have the specialist training, equipment, and platforms necessary for conducting sustained operations in the Arctic have degraded significantly since the end of the Cold War. Regenerating those capabilities would be a costly undertaking and there appears to be little appetite within the Defence community to take that on. More generally, the current direction of travel across the Armed Forces is leading to a situation where Britain has more generic forces that can be adapted to a wide variety of roles, but lacking in the depth of specialisation required to operate in the Arctic (Willett 2013).

Sustained cuts to the Royal Navy have been particularly damaging as, for Britain, the Arctic is primarily a maritime operating space. Britain increasingly has fewer ships and fewer submarines, and those which it does have are being prioritised for use elsewhere in the world. There also remains a persistent threat to the Royal Marines, which are also part of the Royal Navy, that either they or their Arctic warfare training will be scrapped, with their funding allocated for the crewing of the Royal Navy's two new *Queen Elizabeth*-class aircraft carriers instead (Willett 2013). The problem, then, with talking up Britain's security interests in the Arctic, however true those interests may be, is that there remains little political appetite within Defence to open itself up to potential criticism that it is not doing enough.[4]

Externally, the challenge for the British Government is diplomatic. Security concerns in the Arctic, compared to the Middle East for instance, are difficult to discuss, not least because the Arctic states have, over the past decade, gone to great lengths to downplay the potential for conflict in the region. However, although Britain does not have an Arctic topography, it is still vulnerable to threats emanating from the Arctic, especially if the region is destabilised. Thus, while Britain acknowledges that other countries have primary authority over much of the Arctic, it still asserts for itself the right to maintain a watchful eye over what happens along the region's increasingly dynamic borders. That role has been reinforced recently by demands from Parliament that the Government does more to 'protect the Arctic' and guard itself against potential military threats emanating from the region (Parliament 2012; Parliament 2016).

Yet whether or not that sounds reasonable, in the current climate of Arctic diplomacy, it is difficult for non-Arctic states such as Britain to publicly vocalise concerns about security. Too often, statements about security come to be seen in excessively narrow terms as a demand for greater militarisation, or as a kind of self-fulfilling prophecy that will inevitably lead to war (see, for example, Exner-Pirot 2016). Perhaps that is because raising security concerns implicitly involves a judgement about those who are supposed to be the security providers, in this case the A8; an accusation that they are somehow failing in their responsibilities as states to provide security (in the broadest sense, i.e. including protecting people and environments) in their areas of national jurisdiction, and contribute to international peace more generally. Moreover, concepts like the 'domino effect' or 'strategic spillover' have long been used by defence practitioners to legitimise strategies of containment and intervention in fragile, unstable, and vulnerable countries and regions. Further implicit within that is the suggestion that should the A8 fail to maintain adequate security and stability in the Arctic (and it is not clear who is to be judge of that), then it will be the responsibility of others from beyond the region to enforce it themselves.

It is quite evident that that the current British Government does not see imminent potential for conflict in the region, nor does it want that to change. But Britain is still vulnerable to military threats emanating from the Arctic, and the Government has a responsibility to periodically assess the security situation in the region. Nevertheless, because of the political sensitivities of the Arctic states, the Government has shown restraint in how it discusses its Arctic security interests, which is likely why many of its Arctic security reviews have been conducted internally. Whether the recent House of Commons Defence Committee sub-Committee inquiry on 'Defence in the Arctic' changes that approach remains to be seen. The same applies to the re-emerging debate among security scholars about the threat posed by Russian activity in the Arctic to NATO more widely (Olsen 2017). Moreover, the budgetary challenges facing the Ministry of Defence and its subsequent ability to maintain forces for operations in the Arctic seem set to persist. Consequently, while Britain's interests in Arctic defence and security remain high, the political appetite for voicing and debating them remains low.

Enhancing British Science

For centuries, British scientists have made important contributions to understanding the Arctic, including its importance to natural earth systems, as well as facilitating capitalist, colonial, and military expansion into the region. However, in the early twentieth century, the Arctic states invested heavily in national science programmes to help strengthen their sovereign claims to Arctic territories. Britain, meanwhile, following the signing of the Svalbard Treaty in 1920, no longer had any claims in the Arctic, and so successive Governments found little reason to fund a national Arctic science programme equivalent to the British Antarctic Survey (BAS), which helped root Britain's Antarctic claims.

Until the late 1970s, twentieth-century civilian science in the Arctic was therefore mostly led by the universities of Oxford and Cambridge, with the support of learned societies, financial endowments, and private funds. For example, from 1948 to 1992, Walter Brian Harland, a geologist from the University of Cambridge, led several expeditions to Svalbard (known as the Cambridge Spitsbergen Expeditions), with support from countries and oil and gas companies interested in the geology of the Svalbard and Barents continental shelves. The archive of the *Polar Record*, the journal of the Scott Polar Research Institute (SPRI), further shows that British-based scientists had a strong interest in potential Arctic air routes, indigenous cultures, shipping, and fisheries. There had been some close cooperation in fields such as climatology and meteorology between the Armed Services and British-based Arctic scientists during the Second World War and the Cold War, reaffirming a longer tradition of the Royal Navy supporting scientific endeavours of interest to the military. In the 1960s, successive British Governments also provided modest funding for British scientists to take part in the International Biological Programme (1964–1974), which included British contributions to the study of tundra from the British mainland, Greenland, South Georgia, and Signy Island. However, broadly speaking, Government support for British Arctic science was lacking.

Much of the work led by universities was eventually used to inform a seminal textbook entitled *The Circumpolar North*, of which one of SPRI's leading Arctic experts, Terence Armstrong, was a co-author (Armstrong et al. 1978). *The Circumpolar North* was published at a time of growing anxiety in Western capitals over overpopulation, unsustainable natural resource demand, and environmental degradation. This was brought home by events such as the publication of the Club of Rome's 1970 report *The Limits to Growth*, the United Nations (UN) Stockholm Conference on the Human Environment in 1972, and the Middle East oil crisis in 1973. Observing the combined forces of rapid and large-scale post-war industrial economic growth, resource demand, new technologies that seemingly shrunk distances across the globe, climate change, long-distance transboundary pollution, and continuing demands for advanced satellite communications, air transport, and ballistic missiles, the authors concluded that the West's need for scientific knowledge of the Arctic was becoming ever greater (Armstrong et al. 1978).

Yet, aside from an absence of sovereign territory, growing British scientific interest in the Arctic faced two other obstacles. The first was the drag exerted on British polar science by its sovereign interests in Antarctica. To protect these interests, successive governments afforded far greater resources to maintaining a scientific presence in Antarctica, which from 1957 to 1958 had been the focus of the International Geophysical Year (IGY). Scientific activity was a valuable currency in the subsequent international negotiations to establish an 'Antarctic Treaty', signed in 1959, which preserved, but also suspended, all territorial claims to Antarctica (Dodds 2002). In the following decades, successive governments considered the maintenance of a strong scientific presence in Antarctica to be crucial to preserving Britain's influence in the Antarctic Treaty System that subsequently emerged, and which ultimately protects Britain's claim to the British Antarctic Territory.

The other major obstacle was of course the Cold War, during which security considerations largely trumped international scientific cooperation (Archer and Scrivener 2000). The Polish legal scholar Jacek Machowski argued that the emergence of a political and military ice curtain in the Arctic during the Cold War produced an

> atmosphere of mistrust and suspicion, [in which] the institutionalization of multilateral scientific cooperation in the Arctic encountered insurmountable obstacles. (Machowski 1993: 181)[5]

The international cooperation that did exist was mostly based on bilateral arrangements. In Britain's case, that largely meant working with the United States and Canada in Greenland, although the International Biological Programme did provide an example of long-standing and successful international collaboration between East and West.

One exception was Svalbard, where owing to the terms of the Svalbard Treaty, British scientists were able to operate more or less freely. The Cambridge Spitsbergen Expeditions alone ensured that there was a semi-permanent British scientific presence in the Arctic from 1948 onwards. Harland established a summer field base in Ny Ålesund in 1972, which continued to be maintained until 1992. Access to Svalbard was greatly improved in 1975 when an airport was opened in Longyearbyen. As the town opened up to tourism and its infrastructure improved, it became the quickest and easiest access point for British and other European scientists seeking to work in the high latitudes, much more so than Greenland, Alaska, or northern Canada. Soviet islands remained largely off limits.

Increased British Government support for civilian-led Arctic science at its national research centres and universities followed in the 1980s, albeit from a low base. Britain did not own any facilities in the Arctic, and British scientists often relied on overseas funding for their work from sources such as the European Economic Community and the United States Office of Naval Research.

In 1986, with interest in Arctic science cooperation growing in other Arctic and non-Arctic nations, the Natural Environment Research Council (NERC) convened a Polar Sciences Committee (PSC) and hosted a seminar on 'Britain in the Arctic' at the Royal Geographical Society. The PSC's purpose was to assess how scientific investigations of both polar regions might help Britain respond to the latest environmental, technological, and commercial challenges but it also found that Britain lacked a mechanism for coordinating British Arctic science equivalent to the function performed by the BAS in Antarctica (NERC 1989). The need for Britain to have its own national Arctic science programme became more pressing in June 1986, after a meeting of the Scientific Committee on Antarctic Research (SCAR)[6] where informal talks were held on the possibility of establishing a permanent international scientific body for the Arctic. As with SCAR, only those countries able to make a significant scientific contribution would be invited to participate. Meanwhile, at the seminar, British Arctic scientists called on NERC to fund the opening of a national research station in the Arctic.

The PSC went on to establish a working group to review what British science in the Arctic had achieved so far and identify future priorities (NERC 1989). It concluded that although British-based scientists were active in a range of science programmes relevant to the Arctic, their activity was small scale and fragmented, especially compared to British Antarctic research. Without a national body like BAS, through which to channel scientific interest and activity, the science was ad hoc and disparate, and largely relied on funding from sources other than the British Government. Between 1987 and 1988, NERC expenditure on Arctic research was just over £1 million, compared to around £18.5 million on Antarctic research (Miller 1996).[7]

The PSC advised NERC to prepare a 'Strategy for British Research in the Arctic' to set out Britain's core Arctic science interests, and how they should be pursued (NERC 1989). That strategy was eventually published in 1989, a year before the International Arctic Science Committee (IASC) was formally established. However, despite compelling reasons—relating to British military, commercial, and diplomatic interests—for enhancing British Arctic science, NERC stopped short of calling for the creation of a major national Arctic research centre. It argued that SPRI already occupied 'a special position as an academic centre of expertise in polar research and training … [which] functions successfully as a national centre', and that with proper British Government support (i.e. through guaranteed long-term funding), it could continue to act as the institutional base for the national research effort in the Arctic (NERC 1989).[8]

In 1991, NERC established a British scientific research station in Ny Ålesund to be part of the international scientific community that had started to emerge there following Mikhail Gorbachev's Murmansk speech in 1987. Faced with a rapid expansion of scientific activity on the island, Norway decided that the best way to protect the environment and reduce pressure on rescue services was to create an international centre for research in Ny Ålesund. The move was strongly supported by the IASC, but there was perhaps also a geopolitical element involved as a strong international presence on the archipelago would act as a permanent deterrent to Soviet ambitions there (there are still two Russian settlements on the archipelago). Elsewhere, NERC determined that Britain's new dual-purpose logistics and research vessel, the RSS *James Clark Ross*, should be made available to support Arctic as well as Antarctic research. Together, station and ship created, for the first time, an ongoing seasonal British logistical platform in the Arctic from which British scientists could operate, while at the same time firming up Britain's claim to be a major player in the Arctic science community.

Despite NERC's strong statement of British interest in contributing to international Arctic science, until a decade ago British Government support for British Arctic science was still poorly coordinated and conducted in a piecemeal fashion. Money in theory was awarded on the basis of the quality of the science proposed, rather than because they were Arctic studies per se. While there were obvious justifications for that approach, it also worked against the possibility of maintaining a long-term strategic scientific presence in the region. When funding ran out after a few years, whatever presence British scientists had established in the Arctic was typically withdrawn, preventing the accumulation of long-term time series of data so critical for monitoring changes in the environment.

Notwithstanding major individual contributions to international Arctic science, this bottom-up, disparate, and uncoordinated approach in turn made it harder for governments to use British science to leverage influence in emerging intergovernmental bodies like IASC, and later the Arctic Environmental Protection Strategy (AEPS) and the Arctic Council. Britain's subsequent involvement in international Arctic science, focused as it was on specific British interests rather than a broader ambition to contribute across the full spectrum of Arctic science, was therefore limited. While British scientists made important contributions, including as lead authors, to the AEPS/Arctic Council working groups such as the Arctic Monitoring and Assessment Programme (which at the time was particularly focussed on the spread of radioactive contaminants from non-Arctic sites, including Britains Sellafield nuclear fuel reprocessing plant), and the Conservation of Arctic Flora and Fauna (where British scientists contributed to studies of migratory species that visited both Britain and the Arctic), its presence in other working groups was minimal.

The same approach of explicitly rooting Arctic science interests in national concerns was evident in the early 2000s when British scientists took part in the Arctic Climate Impact Assessment (ACIA). Under Tony Blair (Prime Minister, 1997–2007), Britain emerged as a leader of international efforts to understand and address climate change, while domestically the Government was driving through the world's first-ever domestic climate change legislation. When a British anthropologist, Mark Nuttall, was invited by the A8 to serve as a lead author for ACIA, the Foreign Office agreed to fund his participation.[9] That was unusual as the Foreign Office did not normally fund scientific research, but according to an official who was there at the time, the Polar Regions Department was keen to use the opportunity to demonstrate to the rest of Government that engaging with the Arctic Council on climate science could add to Britain's broader efforts in international leadership on climate change, especially since ACIA's findings were of global importance.[10]

That determination to lead on climate change, and the leading contribution made by British scientists during ACIA (and later, their leading roles in the Polar Chapter of the Fourth Assessment Report of the UN Intergovernmental Panel on Climate Change [IPCC], which drew strongly on ACIA), was instrumental in driving Britain's subsequent involvement in the planning and management of the fourth International Polar Year (IPY) from 2007 to 2009. Following the publication of ACIA, NERC and BAS increasingly accepted that the Arctic was where the major impacts of climate change were being felt. Critically, from a British perspective, the IPY was not an Arctic Council-led programme. Rather, it was sponsored by the International Council for Science (ICSU) and the World Meteorological Office, two global organisations where no distinctions were made between the Arctic and non-Arctic states on the basis of circumpolarity. Consequently, within those organisations, Britain shared equal status with the Arctic states. The then-director of BAS, Chris Rapley (1997–2007), had close connections with both organisations and was appointed to co-chair the International Planning Group for IPY, along with an American scientist, Robin Bell. From there he was able to lobby successfully for the IPY's International Programme Office to be based at BAS in Cambridge, a move strongly supported by the Foreign Office because it hoped that both BAS and the International Programme Office could then further contribute to Britain's rising profile in the international Arctic science community. From NERC's perspective, it would also help turn BAS, with its world-leading capabilities in polar science and logistics, into a national centre for the study of *both* polar regions.

With the upsurge of British scientific interest in the Arctic, NERC and the Foreign Office decided to look again at what could be done domestically to support the British Arctic science community. In 2007, it asked the Scottish Association for Marine Sciences (SAMS) to conduct the first major review of British Arctic science since 1989. With BAS still predominantly Antarctica-focused, SAMS was a logical choice owing to its strong record for Arctic marine research and growing international profile. Remarkably, SAMS' conclusions were largely the same as those published by NERC in 1989, suggesting little progress had been made in the intervening years. Foremost among them was the need to increase British Arctic science activity to improve understanding of natural earth systems and help Britain maintain its international lead in predicting the local, regional, and global impacts of climate change (Leakey et al. 2008).

NERC responded to the recommendations by establishing an Arctic Office at BAS in 2009, essentially to act as a national coordinating body (Arctic Committee 2015). The Arctic Office was also made responsible for overseeing a new £10 million Arctic research programme dedicated to better predicting the changes under way in the Arctic. The funding from 2010 to 2016 was eventually increased to over £15 million, representing around a third of all government funding for Arctic science between 2000 and 2016, and therefore a substantial step change in government support (HM Government 2013).[11]

In 2015, NERC announced a further £16 million funding for a successor research programme on the changing Arctic Ocean running from 2017 to 2022, with the aim of better understanding the impact of changing ice conditions on fish, whales, and other sea life, as well as the biochemistry underpinning Arctic ecosystems. The programme in particular seeks to utilise and demonstrate British leadership in an important area of Arctic science (i.e. marine systems), as well as cutting-edge technology (i.e. autonomous vehicles and robotics). In line with NERC's latest science strategy, which calls for science to support British business and economic growth, the programme also has a commercial dimension in that it will explore how changes in Arctic marine ecosystems are likely to affect fisheries and tourism, including in local British waters. Additionally, NERC has invested more than £200 million in a new, world-class polar research vessel—the RSS *Sir David Attenborough*—which is due to be in service in 2019. Although the ship will spend most of its time supporting British science in and around Antarctica, NERC has said it will also be used to enhance Britain's scientific presence in the Arctic. However, although the new ship will be significantly more capable, it is effectively replacing the RSS *James Clark Ross* and RSS *Ernst Shackleton*, and so eventually the number of ships NERC has available for polar science will drop to just one.

The upturn in NERC support for Arctic science is intended to show that while Britain is not, and probably never will be, the leading player in the Arctic in terms of overall science activity (even if it is a global leader in some specific areas like the cryosphere), as a major science nation it can still add to what other nations are doing. That might include, for example, the application of new technologies (e.g. marine robotics, satellites), climate and other predictive modelling, and using science to benefit the commercial sector. The core objective of the British approach to Arctic science is to improve capabilities to predict the regional and global consequences of current and future environmental change, especially in marine systems. NERC's investment is a signal of wanting a more strategic and long-term approach that builds up a British-led observational capability in the marine Arctic for long-term monitoring purposes—effectively playing on British strengths in technology, marine science, and polar logistics. Moreover, the focus on observing and predicting changes is fundamental for providing the science base that feeds other assessments, such as the potential for new shipping routes or fisheries, which other nations might lead on. In short, it is about carving out an important role for Britain at the base of the international Arctic science pyramid, and to then use science capability as leverage for maintaining or increasing Britain's voice in Arctic affairs, without touching on the sensitivities of the A8, and in particular their concerns about sovereignty and security.

Both the creation of an A8 grouping within IASC and the 2017 A8 agreement on scientific cooperation work against the principle of inclusivity on which international scientific activity depends by reasserting geopolitical boundaries between the A8 and the rest. Science is therefore a particularly valuable tool for Britain in the Arctic because it tests the topographies imposed by circumpolarity. That is perhaps why in 2011 the Foreign Office described British science and technology as 'the motor of British activity in the region' (Foreign Office 2011). Within the Arctic Council itself, there are several working definitions of what constitutes the

Arctic. In addition to the Arctic Circle, other ways of marking the Arctic's southern boundaries include the 10°C July isotherm, the treeline, the presence of continuous permafrost, the presence of permanent sea ice, and the marine boundary between cool and warm waters.[12] Defined in these terms, the southernmost limit of the 'Arctic' can be as low as 50°N or as high as 80°N as you travel around the world. Complicating that wavy pattern still further is the fact that increasing climatic and environmental changes—altering, for example, terrestrial and marine temperatures, vegetation cover, and the presence of ice and permafrost—mean that nearly all of these boundaries are becoming more dynamic.

Thus, while the imagined geopolitical boundaries imposed by circumpolarity are strained topographically by dynamism in natural systems, scientific research helps Britain and others to identify new topologies physically connecting events in the Arctic to events elsewhere in the world (China, for example, has recently started to link major air pollution events in its cities to altered weather patterns driven by changes in the Arctic). In Britain, the unusually cold winters experienced in 2009/10 and 2010/11 has led to several scientific studies about whether warming in the Arctic is behind changes in weather in the mid-latitudes (Overland et al. 2016). In the past, the inter-war and post-war expeditions largely led by the universities of Cambridge and Oxford were similarly concerned with connecting Britain and the Arctic through fields as diverse as geology, meteorology, ecology, and ornithology (for example, parts of Scotland were identified as geologically sub-Arctic, while Britain and the Arctic were found to 'share' migratory populations of birds and fish). Scientific investigations can therefore complicate geopolitical claims that Britain and the Arctic are discreet entities.

As scientists and Foreign Office officials draw attention to these shared topologies, they are helping to drive the emergence of narratives about a more 'Global Arctic' (see discussion in Chap. 3), in which the Arctic is increasingly regarded as being intimately connected to the rest of the world through earth systems, and myths of Arctic 'exceptionalism' are expunged. That in turn helps Britain and other non-Arctic states to legitimise their scientific presence in the Arctic, which manifests, for example, in the establishment of scientific research stations on Svalbard, or demands to be included in the work of the Arctic Council and, in particular, conversations about how Arctic changes should be managed. NERC and the Polar Regions Department have come to realise over the past decade or so that they have a major stake in ensuring that British-based scientists are not only active in Arctic research, but that they are also heard. Here, the quality of scientists and scientific institutions at Britain's disposal creates an enormous potential to connect with the Arctic, which in turn would add further weight to Britain's presence on what is generally regarded as the periphery in circumpolar terms (as the strength of Britain's military does).

Yet while the British Government has increased funding for Arctic science, there are still many individuals and institutions that rely on other sourcessuch as the European Union (EU), which has invested in several large-scale international scientific programmes, and the private sector.[13] Interestingly, virtually all of that funding has been directed towards the natural sciences, rather than the social and political sciences, or economic issues. Perhaps funding a comparably large national programme on social sciences or humanities would be seen to touch too closely on the interests and affairs of Arctic states, inviting suggestions of British interference-However, by igoring its social and political scientists, Britain's is risking its ability to help construct and challenge emerging narratives about what kinds of Arctic futures are desirable.

It remains to be seen whether the upscaling of scientific activity seen over recent years is actually helping Britain increase its influence at the Arctic Council, or build stronger links with the rest of the international Arctic science community, or even make money as pressure grows for NERC to commercialise its research outputs. Between 2006 and 2015 Britain ranked fourth in terms of its output of scientific papers on the Arctic, sixth for impact, and first among non-Arctic states in terms of both number of Arctic research projects and funding (Aksnes et al. 2016; Osipov et al. 2016). However, the participation of British scientists in meetings organised by the Arctic Council's various working groups has been limited, especially when those meetings are held in remote parts of the Arctic (where the participation of one British scientist can easily cost upward of £5000), and when there are no clear British interests at stake. More generally, British Arctic science has tended to be more insular and less prominent at international science meetings compared to other Arctic and non-Arctic countries. Such ad hoc participation threatens overall to reduce Britain's presence and influence in international scientific bodies, as well as at the Arctic Council, where new players (notably China, South Korea, and Japan) appear increasingly prepared to connect and contribute across a wide spectrum of projects.

Facilitating British Industry

Commerce has driven English/British interest in the Arctic since at least the fifteenth century. As Britain's commercial interests have evolved, so too have the drivers of the country's economic interest in the Arctic. The British whaling industry, which was more or less active in the Arctic for around 300 years, went into terminal decline in the late nineteenth century as whale oil was substituted for crude oil. By the late 1920s, Britain's coal mines on Svalbard were closed or sold off after international coal prices fell. Britain's distant water fishing fleet, operating mainly out of Grimsby, Hull, and Fleetwood, which had long trawled the waters around Iceland and Greenland, and travelled deep into the Barents Sea, collapsed in the 1970s. But new commercial inroads into the Arctic also appeared. In the 1920s and 1930s, British interest grew in the possibility of establishing commercial air routes across the Arctic between Britain and North America. Following a post-war US Government-led survey of Alaska's hydrocarbon resources which ended in 1953, British Petroleum, which until 1977 was majority-owned by the British Government, was one of the first companies to explore for oil on the North Slope, with production started at Prudhoe Bay in 1968 (Emmerson 2010).

Today, the Arctic is on the verge of a dramatic state change. Much of the Arctic Ocean is changing from being permanently ice-covered into being seasonally ice-free. Many scientists project that the first ice-free summer across the Arctic Ocean will happen before the end of the current century. George Shultz (US Secretary of State, 1982–1989), in an address to congressional policymakers in 2013, likened that transformation to the creation of a new ocean (PSA 2013). Guggenheim, one of the world's largest investment firms, suggest this new ocean could create a $1 trillion economy if all public and private investments in infrastructure currently planned for the next 15 years are delivered. That compares with an existing annual economy in the Arctic of more than $450 billion, and annual demand of $4.5 trillion investment in infrastructure globally (Roston 2016).[14]

These figures should of course be treated with some caution. Ambitious infrastructure projects often produce nothing more than stranded assets. However, they are indicative of growing commercial excitement about the Arctic around the world over the past decade. While several projections about the potential for trans-Arctic shipping,[15] and the development of oil and gas resources,[16] have been overstated by hyperbolic media coverage, commercial interest in the Arctic is growing across other sectors as well, including mining, fisheries, tourism, and telecommunications. In response, in 2010, the World Economic Forum established a Global Agenda Council on the Arctic that devised an Arctic Investment Protocol to guide sustainable investment decisions. In 2015, the A8 announced the creation of an Arctic Economic Council, with responsibility for fostering closer links between Arctic governments and the wider circumpolar business community. Arctic Circle, the annual international Arctic conference, which has convened in Reykjavik every year since 2013, was also started partly in response to growing global commercial interest in the Arctic (Webb 2013).

Contemporary British commercial interest in the Arctic, however, remains unclear and uncertain. There are of course several sectors which *could* be interested, including oil, gas, metals and mining, shipping, offshore renewables, equipment and service supply, engineering, fisheries, tourism, legal and financial services, telecommunications, and research and development. British companies are active, and in some cases, world-leading, across many of those sectors, but that in itself is no guarantee that Britain is going to benefit economically from the Arctic. As the Conservative–Liberal Democrat Coalition Government (2010–2015) made clear in its 2013 Arctic Policy Framework, in the absence of state-owned enterprises (SOEs), the authority to decide whether to invest in an Arctic project rests with the private sector (HM Government 2013).

That said, successive British governments since at least 2010 have shown greater interest in fostering and facilitating British business interests in the Arctic as part of a broader national prosperity strategy that started to emerge more than a decade ago under Tony Blair's premiership. After all, even without a government stake, British-based businesses active in the Arctic still pay taxes in Britain, and dividends to British investors, including pension funds. Greater economic development in the Arctic in key sectors such as food and energy could also benefit Britain by diversifying trade networks and stabilising prices on international markets. And while the Arctic has not been explicitly referenced in papers and speeches related to Britain's prosperity strategy—something which is not unusual as international trade relations generally occur between countries and/or multilateral blocs, as opposed to regions—the general principles of the strategy have been increasingly evident in Britain's engagement with Arctic states, especially Russia and the Nordic countries.

In the most high-profile example, the Coalition Government supported BP (a wholly private entity since 1987) during its attempt to secure a £10 billion deal with the Russian state-backed oil giant Rosneft in 2011. The Government did so for several reasons, including its commitment to facilitate overseas market access for British businesses, its interest in supporting the search for new oil and gas fields to help diversify global market supply (discussed below), and the importance of BP to Britain's economy.[17] The 2010 Deepwater Horizon explosion and oil spill in the Gulf of Mexico had cost BP around $62 billion, and a recovery was badly needed. BP's previous joint venture with Russia (TNK–BP) generated around $19 billion of net dividends for BP's shareholders between 2003 and 2012 so the deal with Rosneft was particularly attractive (BP 2017).

The main impact of Britain's prosperity agenda in the Arctic though has been achieved through cooperation with Nordic countries. In 2011, the British Government signed a Bilateral and Global Partnership with Norway, which among other things reaffirmed the importance of bilateral trade between the two countries (at the time worth around £18 billion). In 2011, David Cameron (Prime Minister, 2010–2016) also hosted the first-ever UK–Nordic–Baltic Summit in London at which he called on the countries involved to 'form an alliance of common interests' and become an 'avant-garde' for economic growth in Europe (Prime Minister's Office 2011). That forum, now known as the Northern Future Forum, continues to meet annually for informal discussions about shared economic interests. In 2012, the Government signed a MOU with Iceland aimed at paving the way for geothermal power from Iceland's volcanoes to supply electricity to Britain through undersea cables. In 2013, a joint UK Trade & Industry (UKTI)/Council of British Chambers of Commerce in Europe report observed in 2013 that the Nordic and Baltic regions had become increasingly attractive to British exporters, especially small- and medium-sized enterprises. UKTI also pointed out several investment opportunities for British businesses emerging in the Nordic countries relating to hydrocarbon development, shipping, mining, infrastructure, and tourism (UK Trade and Investment 2013; see also Gaia 2017).

Successive Governments since 2010 have further highlighted the important role that the City of London's financial and legal services sector could play in support of commercial activity in the Arctic. As a global centre for financial, insurance, and legal services, the City of London has been in a strong position to gain commercially from expanding commercial activity. Arctic projects are capital-intensive and the City is potentially a major source of investment capital. Moreover, the City is world-leading when it comes to underwriting specialist activities in the shipping, energy, and mining sectors, which includes projects being developed in the Arctic. Hundreds of millions of dollars are paid in premiums to London-based underwriting firms for marine insurance to cover oil spills, wreck removal, hulls, equipment, passengers, and crews. Those firms also pay out for losses. Lloyd's Register, which has provided ice classification to the world's ice-going fleet since 1977, has classed around 25% of the world's existing icebreaking fleet (putting it second only to the Russian Register), and is closely involved with fleet renewal projects in Russia and Canada (Hindley 2015). The UK Protection & Indemnity Club has been ensuring risks arising from Arctic trading by offering cover to fleets based in Northern and Far East Russia. Law firms are involved in providing advice and drafting international business contracts (especially international mining contracts), as well as dispute resolution and arbitration, much of which is rooted in English law (40% of governing law in all global corporate arbitrations is English law) and subject to courts based in London (Kingston 2014).

However, the provision of such services also exposes City-based firms to high levels of risk associated with many Arctic projects, including from rogue operators who might flout basic safety and environmental care regulations. Such risks are primarily driven by the hostile environmental conditions that are prevalent across much of the Arctic, but also by the associated reputational risks that come with supporting economic activity in a region that continues to strike an emotive chord with publics across the world; publics who imagine the Arctic to be in need of protection from those who would seek to sully it (Steinberg et al. 2015). The report produced by Lloyd's of London in 2012, entitled *Arctic Opening: Opportunity and Risk in the High North*, emphasised the need for insurers to take seriously the extensive risks associated with commercial operations in much of the Arctic in order to avoid both large payouts and reputational damage; two issues which had been brought sharply into focus by the 2010 Deep Horizon explosion and oil spill, and again after the cruise ship *Costa Concordia* was wrecked off the coast of Italy (Lloyd's of London 2012). With notions of a 'social license to operate' becoming increasingly important to industry worldwide, the British-based insurance sector has been keen to remind operators that a similar event in the Arctic could bring an end to investments across a range of sectors subsequently deemed too risky to operate in the region (Kingston 2016).

Under existing international law, much of the Arctic falls within the jurisdiction of the A8, and it is largely down to them to determine what regulatory frameworks will apply. However, some decisions cannot be taken in isolation from the rest of the international community and what the British Government can do is seek to promote and influence the development and implementation of best practices for businesses internationally. For example, the UN's International Maritime Organisation (IMO), the headquarters of which are in London, recently led the negotiation of a Polar Code, which came into force in January 2017. The Polar Code amends existing international instruments to cover the design, construction, equipment, crew training, safety measures, search and rescue, and environmental requirements related to operating ships in polar waters. Britain and other non-Arctic states have been closely involved in the negotiations, pushing for the highest possible technical, safety, and environmental standards. The work of firms and markets based in the City, such as Lloyd's of London, Lloyd's Register, and Michael Kingston Associates, has been critical to supporting that effort. The City's influence in that regard was recognised by Sweden, which, in 2013, supported the City-based insurance industry's drafting of an Arctic Marine Best Practice Declaration.

With further support from the Arctic states, Lloyd's Register, in conjunction with the International Association of Classification Societies (also headquartered in London) and other IMO participants, went on to develop an ice regime system for assessing, on an individual basis, the risks facing a ship entering polar waters that would support the implementation of the IMO's Polar Code. That has since provided the basis for the insurance industry to help the Arctic Council establish the Arctic Shipping Best Practices Forum in London, firmly entrenching the influence of the City's insurers over the future of Arctic shipping (Kingston 2016). Given such examples, in which British-based businesses directly influenced the agendas of the A8, it is perhaps unsurprising that successive governments since 2010 have been keen to promote the City as a 'centre of commercial expertise with direct relevance to many industries that are growing in the Arctic' (HM Government 2013: 28).

In addition to supporting Britain's national prosperity strategy, there are also other strategic reasons for Britain to promote commercial activity in the Arctic. Among them are concerns about energy security and food security. For instance, in 2003, the British Government published a white paper setting out its long-term energy strategy in response to the emerging challenges posed by climate change (and the need to wean the world of fossil fuels), the implications of reduced oil and gas production from British sectors of the North Sea, and the decline of British coal production (Department for Transport 2003). The challenge posed by the falling commercial viability of continuing to develop Britain's declining reserves of indigenous fossil fuels was of particular concern. The Government estimated that by 2020, Britain could be reliant on imported energy to meet three quarters of its total primary energy (i.e. raw inputs) needs, a dramatic turnaround after being a major oil and gas exporter during the 1980s and 1990s (ibid.).

Subsequent white papers continued to reflect the Government's anxieties over energy security. In 2007, the Labour Government warned that Britain was becoming increasingly dependent on imported fuel at a time when the world's remaining oil and gas reserves were largely thought to be concentrated in the Middle East, North Africa, Russia, and Central Asia; regions generally regarded as vulnerable to resource nationalism (Department of Trade and Industry 2007). Meanwhile, privately owned energy majors were increasingly pushed towards harsher environments and unconventional means in their search for new oil and gas fields. Both developments threatened to destabilise the open markets on which Britain has relied to meet its energy import needs. Consequently, by 2009, Britain's long-term energy security had been reframed as an issue of national security (Cabinet Office 2009; Wicks 2009).

Since 2003, successive governments have responded by promoting open, competitive, and stable global markets in order to maintain security and affordability of supply. Around the same time, it was suggested at a meeting of the Foundation for Science & Technology, a London-based policy forum, that the exploitation of Arctic oil and gas resources could become an important part of Britain's long-term energy security strategy (Goodman 2003). BP, already a long-standing operator on the North Slope in Alaska, had just announced its £6.7 billion deal with a Russian oil company, TNK, to open up new opportunities in the Russian Arctic, while Shell was gaining experience of Arctic-like operations in the ice-infested waters off Sakhalin in the Russian Far East. Arctic oil and gas development, especially where led by international energy companies, was subsequently seen as part of the solution to Britain's long-term energy security. That was highlighted in an internal Foreign Office paper in 2008, which stated:

> [Arctic] resources could contribute to enhancing the UK's security of energy supply, so we should seek to guarantee that the UK can benefit from future Arctic mineral resources. (Foreign Office 2009: 4)

The Department of Energy and Climate Change (DECC) maintained a similar position, and argued repeatedly that Britain could not discount the contribution that the development of hydrocarbon reserves in the Arctic could make to energy security in the coming decades (Environmental Audit Committee 2012).

In the case of Russia, that manifested in the support the British Government showed for the BP–Rosneft deal in 2010. However, the subsequent deterioration of relations between Britain (and the West more broadly) and Russia after the latter's annexation of Crimea in 2014, and the impact of EU and US economic sanctions targeting the development of new offshore oil and gas projects in the Russian Arctic, has prevented further progress (although BP continues to profit from existing operations). Even so, a gas contract signed between Centrica (Britain's largest energy supplier) and Gazprom in Russia in 2015 means that through to 2020, Britain could be reliant on Russia's top natural gas producer (which is developing a massive gas field off the Yamal peninsula in the Russian Arctic) to supply roughly 9% of its gas imports (Schaps 2015).

Meanwhile, Britain's energy relationship with Norway has formed the basis of an increasingly expansive international partnership also encompassing defence, science, and other shared economic interests. Oil and gas imports from Norway grew substantially after Britain became a net importer of energy in 2004.[18] The Norwegian oil and gas industry is now looking to go beyond its maturing fields in the Norwegian Sea and North Sea, and start developing new finds in its offshore Arctic. With demand for gas in Britain growing, and the preponderance of existing infrastructure (i.e. pipelines) connecting Britain to gas flows from Norway and Russia, it is therefore difficult to escape the fact that Britain's energy security will be closely connected to developments of Arctic prospects for the foreseeable future, at least until alternative sources of energy (whether that be from indigenous shale reserves, liquefied natural gas [LNG] from the Middle East and North America, or renewables) are further developed.

Britain's other significant strategic commercial interest in the Arctic relates to food security, specifically fisheries. In 2008, the Cabinet Office warned about the challenges of operating in an increasingly globalised marketplace, which was brought home by the world food price crisis between 2007 and 2008 (Cabinet Office 2008). The Government expressed particular concern about the decline in global fish stocks. Britain remains a major importer of its most popular fish (cod and haddock). In 2016, its fish processing sector provided close to 18,000 jobs. In 2014, the sea fish annual industry turnover was about £3.1 billion (Seafish 2016).

As a consequence of inadequate domestic stocks, Britain is increasingly dependent for its needs on other countries adopting sustainable practices to manage their fisheries. Around 70% of all Atlantic cod that ends up in supermarkets around the world comes from the Barents Sea; 95% of the cod sold by British fish and chip shops is also caught in Arctic waters. Iceland, Norway, Canada, Denmark, and the Faroe Islands are all among the top suppliers of fish to Britain. Major British seafood brands (including Birds Eye, Findus and Young's) also rely on fish from the Arctic. So, while Britain no longer has a large distant water fleet directly involved in Arctic fishing, it still has substantial interests in what happens to both current and future fisheries in the Arctic. Quite what that future will be remains uncertain, but some marine policy experts believe that as the sea ice retreats and Arctic waters warm, several fish species are likely to move northwards, leading to much larger takes from Arctic fisheries, including in the international waters of the Central Arctic Ocean (Pan and Huntington 2016: 153).

In 2013, the British Government emphasised the importance of working with the rest of the EU to encourage sustainable management of Arctic fishing and fisheries (HM Government 2013). The Arctic Ocean littoral states regulate the fisheries that fall within their Exclusive Economic Zones (EEZs), but future access to the international waters of the Central Arctic Ocean is currently under consideration in negotiations led by the A5, together with a further five major fishing actors (Iceland, China, South Korea, Japan, and the EU). However, Britain's involvement in the Central Arctic Ocean negotiations has seemingly become more precarious since the Government declared its intent that the country should leave the EU in March 2019. Lacking a sizeable distant water fishing fleet of its own, it is hard to imagine that Britain will be invited to take part in the negotiations as an independent state unless the terms of participation are changed. While Britain may still have contributions to make in terms of science, monitoring and enforcement, and potentially also best practice (in 2016, several British seafood forms were part of an industry group agreement to avoid expanding cod fishing with bottom-trawlers into areas of the Arctic where there has not been regular fishing before), its direct involvement in fisheries negotiations is likely to be limited. Yet owing to its ongoing dependence on fish imports to satisfy domestic demand and commercial interests, Britain's strategic stake in the commercial and sustainable success of existing and future Arctic fisheries will likely remain high.

Other areas of commercial interest in the Arctic abound, and some of those may yet become strategic. In 2016, a Chinese ship travelled from Tianjin to Sheerness via the Northeast Passage to deliver a cargo of wind turbines. If Arctic waterways ever become significant routes for international trade, Britain would likely have a strong strategic interest in ensuring that those routes remain freely navigable. Likewise, increased interest in laying down fibre-optic cables (which could speed up the transfer of data between financial capitals in Europe [including London], North America, and Asia) or energy interconnectors (which could bring renewable energy from the Arctic to Britain) along the Arctic Ocean seafloor could also create new strategic commercial interests for Britain in the region. However, decisions about what kinds of shared commercial topologies are forged between Britain and the Arctic will still, in Britain at least, largely be dominated by private sector interests, and whether they attach enough value to what the region has to offer. Britain's topographical proximity to the Arctic will no doubt play its part in shaping at least some of those connections (energy interconnectors, pipelines, fibre-optic cables, and so on, all build on the potential which that topography offers), but so too will the topologies of economic globalisation (i.e. the capacity for finance, services, people, and material to flow easily around the world), which depend not so much on physical proximity, but rather the kind of proximity engendered by the ability to connect one site of commercial activity to another, something which the City of London, in particular, excels at.

Conclusion

British interest in the Arctic today is connected to contemporary and future challenges emerging in the region, whether those relate to climate change and environmental pollution, commercial activity, or governance, geopolitics, and security. Some of those connections exist because of Britain's Arctic history and its relative topographical proximity to the region. For instance, in the case of the former, among sections of the defence community, present-day concerns about the military threat Russia poses in the Arctic are rooted in memories of the threat that the Soviet Union posed to its allies during the Cold War, and so the recent past comes to act as a guide to the present. Meanwhile, climate scientists look at the distant past to help them understand what an ice-free Arctic could look like, and how that would affect the rest of the world, and so the Arctic comes to be connected with the long-standing interests of British scientists in earth systems. In the case of the latter, Britain's topographical proximity to the Arctic makes it easier to build certain types of material connections to the region that facilitate greater connectedness, such as the pipelines, electricity interconnectors, and fibre-optic cables that facilitate trade. It also makes Britain more vulnerable to material spillover from the region, whether in terms of a spreading oil spill that reaches the North Sea, geopolitical instability that requires an increased military presence in the North, or a severe weather event on the British mainland intensified or made more likely by climatic instability in the Arctic.

But history and topography can also be poor guides to Britain's contemporary role and interests in the Arctic. It does not really matter to the shape of present-day commercial interests in the Arctic that Britain was once a major whaling, sealing, and fishing power that regularly trawled Arctic waters. Nor does it really matter that the City of London is relatively close to the Arctic topographically, compared to other major financial capitals, because the City's importance to the Arctic is not based on its—or event Britain's—physical proximity to the region, but on its ability to connect flows of finance and information around virtually the entire world, almost instantaneously. Likewise, whether the British military has any interest in the Arctic would be far less significant if Britain was not still the largest military power in Europe, nuclear-armed, and one of a handful of countries with strategic submarine forces capable of operating under the ice.

Overall, what really seems to matter, then, is not that Britain is the Arctic's 'nearest neighbour', topographically speaking, or that it has a long history of activity in the Arctic. Rather, what matters is that Britain today retains a vast potential to connect itself with what is happening in the Arctic through a myriad of stakeholder interests broadly relating to defence, science, and commerce. While some of that connectivity is aided by Britain's topographical proximity to the Arctic as well as legacies of historical activity, it is the potential and actual connectedness of contemporary actors, practices, and sites of activity that is determining how much Britain can shape the present and future Arctic. That is because those contemporary actors, practices, and sites of activity give *weight* to Britain, which acts as a pull on the Arctic, testing circumpolar constructs by stretching the Arctic towards Britain and changing perceptions about distances between the two. But they also produce particular patterns of connectivity that vary over time depending on which actors, practices, and sites of activity are most active in creating and maintaining connections at any particular moment. Multiple, and at times, overlapping and conflicting ideas about what kind of place the Arctic is—an abomination, a transit route, a scientific laboratory, a resource frontier, a military theatre, a vulnerable environment, a home to indigenous peoples, and a place for adventure—have therefore varied in prominence throughout Britain's Arctic history, making the Arctic seem more or less distant in the process. Today, the situation is no different as concerns about Russia, energy security, and climate change have variously inspired politicians, military officers, scientists, environmentalists, businesses, and others to demand greater British involvement in Arctic affairs.

Such connections are not easily mapped in topographical terms though. They are essentially invisible at the resolutions of scale required to produce meaningful maps of the Arctic. Instead, to see them, a more abstract act is required, one which foregrounds topology (i.e. connectivity and connectedness) before topography (i.e. physical distances). Just as two people on opposite sides of the world can maintain close relations and a sense of proximity to each other, while being entirely ignorant about their neighbours living next door, so too can different geographical constructs (states, regions, etc.) be rendered more or less proximate to each other by the connections that are made between them.

Consequently, Britain and the Arctic matter to each other today not just because of history and topographical geography, but because extensive connections, broadly produced through activities relating to defence, science, and commerce, continue to exist. The implications, however, are doubled-edged. On the one side, the more connections that Britain creates with the Arctic, for instance, by building up military forces, increasing scientific activity, or promoting new trade links, the more it can implicitly or explicitly influence what happens there. On the other side, if those connections break down, whether deliberately or through neglect, then no amount of history or topographical proximity is going to give Britain any influence over what happens in the region. That is why it matters, to the defence community, whether cuts to the Armed Forces make it impossible for Britain to operate in the Arctic; to commercial stakeholders, whether

British-based firms are able to take advantage of economic opportunities in the region and set examples of best practice; and to scientists, whether sufficient funding is made available to allow British research institutes to play an active role in Arctic science programmes. It is also why it matters when the Arctic states threaten to throw up barriers to greater connectivity through too great an emphasis on upholding circumpolarity, and why the British Government has to be cautious about how it negotiates Britain's present and future connectedness to the region.

Notes

1. The phrase 'post-West' world order was used by Russian Foreign Minister Sergey Lavrov in his speech at the Munich Security Conference in 2017.
2. Author interview with two Ministry of Defence Officials, 1 May 2012.
3. In contrast, another major European power such as Germany, lacking Britain's combination of geography and capability to project force, would be unlikely to exert the same effect.
4. In 2011, for example, a Defence Correspondent at *The Telegraph* was quick to criticise the Government for being 'woefully unprepared for Arctic warfare' after a joint Foreign Office/Ministry of Defence paper on security challenges facing Britain in the Arctic was leaked (Harding 2011).
5. Machowski compares that with Antarctica, 'where a well-defined and implemented international scientific cooperation has been developed early and successfully' following the IGY 1957–1958 (Machowski 1993: 181).
6. That such talks should have occurred at SCAR may sound surprising but during the IGY, the International Council for Science (ICSU) initially discussed a proposal to establish a Scientific Committee on Arctic and Antarctic Research. However, participants in those discussions concluded that given numerous political and strategic obstacles in the Arctic, it was only feasible to establish a body for Antarctica—SCAR (Machowski 1993).
7. Over the next decade, NERC expenditure on the Arctic remained between £1 and £2 million per year, while expenditure on Antarctica was over £20 million (peaking in 1990 at £48 million).
8. Although it also worth noting that representatives from SPRI were closely involved in the writing of the NERC strategy and therefore may well have been protecting their own interests here too.
9. A second British scientist, Terry Callaghan, was also invited to serve as a lead author but was directly funded by Sweden.
10. Author interview with Foreign Office official, 5 October 2011.
11. Although to help put that figure in perspective it is worth noting that the British Antarctic Survey's *annual* operating budget is around £50 million, most of which is funded by the Government and directed towards

Antarctica (although that figure includes the maintenance of logistics and facilities, and training, in addition to research, a small proportion of which happens in the Arctic).

12. Other political and cultural definitions also exist, wherein, for example, the Arctic is defined by the presence of indigenous peoples.
13. The implications of Britain leaving the EU for British participation in such programmes remain uncertain.
14. Guggenheim has broken the $1 trillion figure down roughly as follows: Energy ($193 billion), Mining ($80 billion), Renewable Energy ($61 billion), Rail ($23 billion), Industry ($19 billion), Maritime ($17 billion), Road ($13 billion), Power ($13 billion), Tourism ($5 billion), Aviation ($3.3 billion), Social ($2.4 billion), Telecom ($1.8 billion), and Trade ($1.7 billion).
15. The Northeast and Northwest Passages reportedly offer distance savings of between 17% and 30% for shipping between East Asia, Northern Europe, and northerly parts of North America when compared with traditional routes using the Suez and Panama canals.
16. The widely publicised US Geological Survey report of 2008 claimed that total estimated hydrocarbon resources in the Arctic amount to around 30% of the world's undiscovered gas and 13% of undiscovered oil. Critically, these estimates relate to resources which are technically recoverable. It does not make any assessment of whether they are commercially recoverable.
17. In 2010, £1 of every £7 paid in dividend to UK pension funds by FTSE 100 companies came from BP. In 2009, UK taxes linked to BP totalled £5.8 billion (Reuben 2010).
18. Britain's crude oil imports from Norway peaked at around 67% in 2011 (they still make up roughly 50%). Gas imports from Norway doubled between 2006 and 2008, rising to 55% in 2012 and 61% in 2015.

References

Aksnes, Dag, Igor Osipov, Olga Moskaleva, and Lars Kullerud. 2016. *Arctic Research Publication Trends: A Pilot Study*. University of the Arctic. Accessed July 18, 2017. https://www.elsevier.com/__data/assets/pdf_file/0017/204353/Arctic-Research-Publication-Trends-August-2016.pdf

Archer, Clive, and David Scrivener. 1986. Introduction. In *Northern Waters*, ed. Clive Archer and David Scrivener, 1–10. Totowa: Barnes & Noble.

———. 2000. International Cooperation in the Arctic Environment. In *The Arctic: Environment, People, Policy*, ed. Mark Nuttall and Terry Callaghan, 601–220. Amsterdam: Harwood Academic Publishers.

Arctic Committee. 2015. The Natural Environment Research Council – Written Evidence (ARC 0041). Accessed June 22, 2017. http://data.parliament.uk/

writtenevidence/committeeevidence.svc/evidencedocument/the-arctic-committee/arctic/written/13355.html

Armstrong, Terence, George Rogers, and Graham Rowley. 1978. *The Circumpolar North*. London: Methuen & Co Ltd.

Åtland, Kristian. 2014. Interstate Relations in the Arctic: An Emerging Security Dilemma? *Comparative Strategy* 33: 145–166.

BP. 2017. Working in Russia. *BP*. Accessed June 22, 2017. http://www.bp.com/en_ru/russia/about-bp-in-russia/business.html

Cabinet Office. 2008. *Food Matters: Towards a Strategy for the 21st Century*. London: Cabinet Office.

———. 2009. *The National Security Strategy of the United Kingdom: Update 2009 – Security for the Next Generation*. London: The Stationary Office.

Conley, Heather, and Caroline Rohloff. 2015. *The New Ice Curtain: Russia's Strategic Reach to the Arctic*. Lanham: Rowman & Littlefield.

Department for Transport. 2003. *Our Energy Future: Creating a Low Carbon Economy*. London: The Stationary Office Limited.

Department of Trade and Industry. 2007. *Meeting the Energy Challenge: A White Paper on Energy*. London: The Stationary Office Limited.

Depledge, Duncan, and James Rogers. 2016. Securing the Wider North. *RUSI Newsbrief* 36: 19–20.

Dodds, Klaus. 2002. *Pink Ice: Britain and the South Atlantic Empire*. London: I.B. Tauris.

Emmerson, Charles. 2010. *The Future History of the Arctic*. London: The Bodley Head.

Environmental Audit Committee. 2012. *Protecting the Arctic*. London: The Stationary Office Limited.

Exner-Pirot, Heather. 2016. Northern Expert: Put Up or Shut Up with Your Arctic Conflict Theory. *Alaska Dispatch News*, September 28. Accessed June 22, 2017. https://www.adn.com/arctic/article/northern-expert-proposes-new-rule-put-or-shut-your-arctic-conflict-theory/2015/10/21/

Foreign Office. 2009. *The Arctic: Strategic Issues for the UK*. London: Foreign and Commonwealth Office.

———. 2011. The UK's Engagement in the Arctic. *Foreign and Commonwealth Office*. Accessed May 29, 2012 – *no longer available*.

Fox, Liam. 2008. Speech to the House of Commons. *Parliamentary Debates*, May 8, Commons, vol. 475, col. 894. Accessed June 22, 2017. https://www.publications.parliament.uk/pa/cm200708/cmhansrd/cm080508/debindx/80508-x.htm

Foxall, Andrew. 2014. 'We Have Proved It, the Arctic Is Ours': Resources, Security and Strategy in the Russian Arctic. In *Polar Geopolitics? Knowledges, Resources and Legal Regimes*, ed. Richard Powell and Klaus Dodds, 93–112. Cheltenham: Edward Elgar.

Gaia. 2017. *Mapping Opportunities for the UK in the Arctic: Study Report for the British Embassy in Helsinki*. British Embassy, Finland. Accessed June 23, 2017. https://www.gov.uk/government/uploads/system/uploads/attachment_data/file/591445/Report_Mapping_of_UK_business_opportunities_in_the_Arctic.pdf

Goodman, Dougal. 2003. Opportunities for Prudent Operators. *Foundation for Science & Technology Journal* 18: 14–15.

Harding, Thomas. 2011. Britain 'Woefully Unprepared' for Arctic Warfare. *The Telegraph*, August 10. Accessed August 14, 2017. http://www.telegraph.co.uk/news/uknews/defence/8692295/Britain-woefully-unprepared-for-Arctic-warfare.html

Hindley, Rob. 2015. LR Teams Pioneer Series of Global Icebreakers. *Horizons*, May 20–23.

HM Government. 2013. *Adapting to Change: UK Policy Towards the Arctic*. GOV.UK. Accessed October 26, 2016. https://www.gov.uk/government/uploads/system/uploads/attachment_data/file/251216/Adapting_To_Change_UK_policy_towards_the_Arctic.pdf

Holtsmark, Sven, and Brooke Smith-Windsor, eds. 2009. *Security Prospects in the High North: Geostrategic Thaw or Freeze?* Rome: NATO Defence College.

Hudson, Peter, and Peter Roberts. 2017. The UK and the North Atlantic: A British Military Perspective. In *NATO and the North Atlantic: Revitalising Collective Defence*, ed. John Andreas Olsen, 75–91. London: Royal United Services Institute.

Johnstone, Rachael, and Federica Scarpa. 2016. Little Italy: Seeking a Niche in International Arctic Relations. *Icelandic E-Journal of Nordic and Mediterranean Studies* 11. Accessed February 17, 2017. http://nome.unak.is/wordpress/volume-11-no-1-2016/01_double-blind-peer-reviewed-article/little-italy-seeking-niche-international-arctic-relations/

Jónsson, Hannes. 1982. *Friends in Conflict: The Anglo-Icelandic Cod Wars and the Law of the Sea*. London: Hurst.

Kingston, Michael. 2014. *Realities of Operating in the Arctic: Finance, Legal and Insurance Issues and International Regulation – The Polar Code Developments*. Paper presented at Arctic Circle, Reykjavik, Iceland, November 1.

———. 2016. *Arctic Shipping and the City of London*. Paper presented at the All-Party Parliamentary Group for Polar Regions, London, November 2.

Leakey, Ray, Finlo Cottier, John Howe, Travis Potts, Robert Turnewitsch, and David Meldrum. 2008. *A Review of the Current Status of UK Arctic Research*. Oban: The Scottish Association for Marine Science.

Lloyd's of London. 2012. *Arctic Opening: Opportunity and Risk in the High North*. London: Lloyd's of London.

Machowski, Jacek. 1993. IASC as Legal Framework of International Scientific Cooperation in the Arctic. *Polish Polar Research* 14: 177–207.

Mazo, Jeffrey. 2015. Showing the Flag. *Survival* 57: 241–252.
Mazo, Jeffrey, and Christian Le Miere. 2013. *Arctic Opening: Insecurity and Opportunity*. Abingdon: Routledge.
Miller, Baroness of Hendon. 1996. Written Answers (Lords), vol. 571, col. 64WA–65WA, April 16. Accessed June 22, 2017. http://hansard.millbanksystems.com/written_answers/1996/apr/16/polar-antarctic-and-arctic-research#S5LV0571P0_19960416_LWA_46
Ministry of Defence. 2010. Defence Secretary Launches New Forum of Northern European Countries. *Ministry of Defence*. Accessed June 22, 2017. https://www.gov.uk/government/news/defence-secretary-launches-new-forum-of-northern-european-countries
NERC. 1989. *Britain in the Arctic*. Swindon: Natural Environment Research Council.
Olsen, John Andreas. 2017. Introduction: The Quest for Maritime Supremacy. In *NATO and the North Atlantic: Revitalising Collective Defence*, ed. John Andreas Olsen, 3–7. London: Royal United Services Institute.
Osipov, Igor, Gils Radford, Dag Aksnes, Lars Kullerud, Diane Hirshberg, Peter Skold, Kirsi Latola, Olga Moskaleva, and Aaron Sorensen. 2016. *International Arctic Research: Analyzing Global Funding Trends. A Pilot Report*. University of the Arctic. Accessed July 18, 2017. http://research.uarctic.org/research-area/research-analytics-task-force/publications/
Overland, James, Klaus Dethloff, Jennifer Francis, Richard Hall, Edward Hanna, Seong-Joong Kim, James Screen, Theodore Shepherd, and Timo Vihma. 2016. Nonlinear Response of Mid-Latitude Weather to the Changing Arctic. *Nature Climate Change* 6: 992–999.
Pan, Min, and Henry Huntington. 2016. A Precautionary Approach to Fisheries in the Central Arcitc Ocean: Policy, Science, and China. *Marine Policy* 63: 153–157.
Parliament. 2012. Environmental Audit Committee Publishes Report on Protecting the Arctic. Accessed October 26, 2016. https://www.parliament.uk/business/committees/committees-a-z/commons-select/environmental-audit-committee/news/announcement-of-report-publication1/
———. 2016. Defence in the Arctic Inquiry Launched. Accessed June 22, 2017. https://www.parliament.uk/business/committees/committees-a-z/commons-select/defence-committee/defencesubcommittee/news/defence-in-the-arctic-inquiry-launch-16-17/
Prime Minister's Office. 2011. PM Hosts Nordic Baltic Summit in London. *Prime Minister's Office*. Accessed June 22, 2017. https://www.gov.uk/government/news/pm-hosts-nordic-baltic-summit-in-london
PSA. 2013. George Shultz, Fmr. Sec of State, Addresses Policymakers on Capitol Hill for the First Time in 20 Years. *Partnership for a Secure America*, March 8. Accessed June 22, 2017. http://www.psaonline.org/2013/03/08/george-shultz-addresses-policymakers/

Reuben, Anthony. 2010. Why is BP Important to the UK Economy? *BBC News*, June 10. Accessed 22 June 2017. http://www.bbc.co.uk/news/10282777

Roston, Eric. 2016. The World Has Discovered a $1 Trillion Ocean. *Bloomberg Business*, January 21. Accessed June 22, 2017. https://www.guggenheimpartners.com/GP/media/pdf/Bloomberg_The-World-Has-Discovered-a-1-Trillion-Ocean.pdf

Schaps, Karolin. 2015. UK Gas Import Dependence Grows with Centrica's Russia, Norway deals. *Reuters*, May 13. Accessed June 22, 2017. http://uk.reuters.com/article/uk-centrica-gas-deals-idUKKBN0NY1FH20150513

Seafish. 2016. Processing Sector Statistics. *Seafish*. Accessed June 22, 2017. http://www.seafish.org/research-economics/industry-economics/processing-sector-statistics

Staveley, William. 1988. An Overview of British Defence Policy in the North. In *Britain and NATO's Northern Flank*, ed. Geoffrey Till, 65–73. London: Macmillan Press.

Steinberg, Philip, Jeremy Tasch, Hannes Gerhardt, Adam Keul, and Elizabeth A. Nyman. 2015. *Contesting the Arctic: Politics and Imaginaries in the Circumpolar North*. London: I.B. Tauris.

Taylor, Ian. 2008. Speech to the House of Commons. *Parliamentary Debates*, May 8, Commons, vol. 471, col. 386. Accessed August 7, 2017. https://publications.parliament.uk/pa/cm200708/cmhansrd/cm080130/debtext/80130-0015.htm

Taylor, Ann. 2009. Speech to the Joint NATO/Icelandic Government Conference (Security Prospects in High North), Reykjavik, Iceland. *Ministry of Defence*. January 29. Accessed June 22, 2017. http://webarchive.nationalarchives.gov.uk/+/http:/www.mod.uk/DefenceInternet/AboutDefence/People/Speeches/MinISD/20090129JointNatoicelandicGovernmentConferencesecurityProspectsInHighNorthReykjavicIceland.htm

UK Trade and Investment. 2013. *Think Nordic & Baltic: Business Opportunities on Your Doorstep*. UK Trade and Investment. Accessed June 23, 2017. https://www.gov.uk/government/publications/think-nordic-baltic

Webb, Robert. 2013. Iceland President Sounds Climate Alarm Demanding Global Attention, Action at NPC Luncheon. *The National Press Club*, April 15. Accessed January 6, 2017. https://www.press.org/news-multimedia/news/iceland-president-sounds-climate-alarm-demanding-global-attention-and-action

Wicks, Malcolm. 2009. *Energy Security: A National Challenge in a Changing World*. London: Department of Energy & Climate Change.

Willett, Lee. 2013. Afterword: A United Kingdom Perspective on the Role of Navies in Delivering Arctic Security. In *Arctic Security in an Age of Climate Change*, ed. James Kraska, 281–298. Cambridge: Cambridge University Press.

CHAPTER 5

To Strategise in the Arctic, or Not?

Abstract As British interest in the Arctic has heightened over the past decade, civil servants, politicians, scientists, civil society, and others have debated whether the British governments should devise and articulate a detailed strategy setting out how Britain will pursue its interests in the Arctic over the coming years. The British Government has so far resisted, but in 2013 did publish an Arctic Policy Framework. This chapter shows that although the Arctic Policy Framework cannot be considered a strategy, it is intended to have strategic effects—namely, to placate domestic critics calling for the Government to do more to support British stakeholders in the Arctic, and reassure the Arctic states that Britain is not a threat to the geopolitical status quo in the region. However, it remains a precarious framework.

Keywords Britain and the Arctic • Strategy • Arctic Policy Framework • Arctic geopolitics

On 10 March 2008, the Scottish Association of Marine Science (SAMS) hosted a meeting for 39 representatives from the British Government (including officials from the Foreign Office, Ministry of Defence, Department for Environment Food and Rural Affairs, as well as several other agencies and advisory bodies), national research centres, universities, and businesses convened for a three-day workshop in Oban, Scotland. The purpose of the workshop was to discuss recent geopolitical developments

© The Author(s) 2018

D. Depledge, *Britain and the Arctic*,
https://doi.org/10.1007/978-3-319-69293-7_5

in the Arctic, and the opportunities and challenges those developments posed to Britain.

Before the workshop, SAMS had been commissioned by the Polar Regions Department in the Foreign Office to produce a report on the state of British Arctic science. The timing of both the report and the workshop was unlikely to have been coincidental. British scientists had already made impactful contributions to the Arctic Climate Impact Assessment (ACIA) and the Fourth Assessment Report of the Intergovernmental Panel on Climate Change (IPCC), and the third International Polar Year (IPY 2007–2009), for which Britain was hosting the International Programme Office, was in full swing. The Natural Environment Research Council (NERC) had also recently identified understanding polar change and its global consequences as a priority (NERC 2007). Meanwhile, in the international media, the Arctic had become a hot topic following the new record sea ice low in 2007, speculation about oil and gas resources, as well as future shipping lanes, and the intrigue surrounding the Russia-backed *Arktika* expedition to the North Pole. As the Executive Summary of the resultant report from the Oban workshop noted, the moment was ripe 'to discuss the key issues relating to UK interests in the Arctic' (GBSC 2008: 3).

The workshop addressed a variety of British interests in the Arctic including defence, energy security, Arctic science, climate science, conservation, biodiversity, industry, shipping, fisheries, and governance, with the principal aim being to 'address the key issues and challenges the UK faces in the Arctic and to have in-depth discussion surrounding the UK policy approaches to the region' (GBSC 2008: 3). Among several questions posed for the event, the organisers asked, 'Should we have an overarching UK–Arctic policy or strategy?', 'How should the UK engage on matters of Arctic Governance?', 'What sort of Governance framework would best deliver to the priorities of the UK?' (GBSC 2008: 3).[1] The workshop went on to reflect on the very nature of Britain's relationship with the Arctic, and how it should be taken forward in the context of growing domestic and international interest in the region. As such, it is likely to have been the most significant and sustained discussion of Britain's Arctic interests since the mid-1990s, when the activities of the National Arctic Research Forum and the Northern Waters and Arctic Study Group had tailed off (see Chap. 2).

The outcome of that workshop is the jumping-off point for this chapter. As the workshop report concluded, the participants' main conclusion was that there needed to be 'better coordination across the whole of the UK sphere of [Arctic] interests'. However, the report stopped short of calling for the Government to prepare an overarching Arctic policy or strategy. This chapter offers an explanation of why. It then goes on to analyse why, in 2013, the Conservative–Liberal Democrat Coalition Government (2010–2015) reconsidered its approach and published an 'Arctic Policy Framework' (HM Government 2013). From there, the chapter enquires into the *strategic* nature of the Arctic Policy Framework, and why successive British governments since have apparently been so reluctant to go further and publish an Arctic Strategy, as all the Arctic states have done. In doing so the chapter asks: is the Arctic Policy Framework a strategy in all but name? And what does it say about how the Government is trying to frame Britain's relationship to the Arctic today?

The 'Strategic Gap'

In his sweeping history of strategy, Sir Lawrence Freedman, one of the world's leading authorities on war and international politics, states that 'strategy' is

> about getting more out of a situation than the starting balance of power would suggest. It is the art of creating power. (Freedman 2013: xii)

While power is a notoriously tricky term to define, what Freedman seems to be saying is that strategies are used to turn a situation in one's own favour while being responsive to the fact that others are trying to do the same. As Freedman explains in his study, strategy is needed when 'situations are uncertain, unstable and unpredictable' (Freedman 2013: 612). Strategies differ from plans because they are responsive to the potential for outcomes to be contested. So, while plans are directed towards realising a single, stable vision of the future through the means that are available, strategy is about making sense of a range of possible futures, and the extent to which desirable outcomes can be brought about given the means to hand.

The workshop in Oban raised the prospect of the British Government developing an overarching strategy for the Arctic. What has been happening in the Arctic over the past decade has certainly invited strategic consideration from successive British governments, whether because of climate change and energy security, as it was under Labour, or energy security, trade, and relations with Russia, as it has been under recent governments led by the Conservatives. As the workshop report emphasised:

> The speed of climate change in the Arctic and the associated impacts and opportunities mean that 'business as usual', with respect to the way the UK interacts with the region (both politically and commercially), is unlikely to be a sustainable or viable approach. (GBSC 2008: 2)

The report goes on to highlight what interests are at stake for Britain. These relate primarily to goals such as protecting the Arctic environment, exploiting resources sustainably, understanding the Arctic environment and the effects of climate change, and having influence among the Arctic states. Several uncertainties about the Arctic are also mentioned, such as whether there would be a level playing field for British business, how new fisheries would be managed, whether new shipping routes would emerge, how future scientific activity would be funded, what would be the effects of climate change, and how Britain should engage with Arctic states and governance structures. The need for the Government to 'create power' in the Arctic to drive forward its interests is recognised implicitly in the report's call for Britain to 'punch above its weight at the Arctic Council' (GBSC 2008: 5).

Yet while the contents of the report provide a firm basis for beginning a discussion about what kind of strategy Britain might adopt to manage uncertainties and push forward its interests in the Arctic, the report itself stopped short of calling for the British Government to produce a formal strategy in the form of a public white paper. There are several reasons why that might have been the case. The first was that there was no clear departmental lead for Arctic policy within the Government. Changes in the Arctic impact several policies and issues, responsibility and accountability for which are distributed across numerous departments and agencies.[2] In the mid-1990s, the Polar Regions Department had only become responsible for representing Britain in certain Arctic international forums, principally the Arctic Environmental Protection Strategy (AEPS) and, later, the Arctic Council. However, those forums did not (and still do not) address all issues pertaining to the Arctic. To the contrary, the Arctic is governed by an array of international agreements, policy declarations, legal instruments, and subregional/subnational organisations, much of which falls outside the limited remit of the Polar Regions Department, and even beyond the remit of the Foreign Office.[3]

The second reason was that British Government officials have been resistant to framing the Arctic as a homogenous region in which to act. After all, as noted in Chap. 3, geopolitically, the Arctic comprises eight Arctic states surrounding an ocean, which is subject to the law of the sea. Within that, there are areas that are ice-covered for most of the year (i.e. the 'white Arctic') and areas which are almost entirely ice-free (i.e. the 'blue Arctic'). Within the Arctic states, there are various degrees of devolution, granting rights of self-governance to indigenous peoples (such as the Inuit in Canada) and nations (i.e. Greenland and the Faeroe Islands). Svalbard is governed by Norway under the terms of the 1920 Svalbard Treaty. Fish, oil, gas, and other resources are unevenly distributed, as are the impacts of environmental pollution and climatic changes. Each Arctic state enforces different legal and fiscal regimes. Some parts of the Arctic (such as Russia's Kola Peninsula) are heavily militarised. All of the Arctic states except Russia are part of North Atlantic Treaty Organization (NATO) or the European Union (EU). An overarching mosaic of different governance regimes covers the entire region. That heterogeneity called into question the utility of an overarching strategy for the region.[4]

Related to that, British Government officials were also resistant to framing the Arctic as an 'exceptional' region. Arctic 'exceptionalism' has been a common refrain in Arctic policy circles since Mikhail Gorbachev's 1987 Murmansk speech, in which he called for the region to be turned into a zone of peace set aside from strategic power struggles (Käpylä and Mikkola 2015). However, it is a concept that successive British governments have in some ways struggled to relate to as its interests in the Arctic are defined less by the region's exceptionalism and more by how the Arctic is connected with a network of broader national interests relating to shipping, energy security, environmental protection, and so on. In that regard, an overarching strategy for the Arctic, addressing Arctic shipping, Arctic energy, and Arctic environmental concerns, was assessed to be of less value than working to ensure that British interests in 'global' shipping, 'global' energy security, and 'global environmental protection' were responsive to the changes occurring in the Arctic.[5,6]

The third reason was that British Government officials (especially in the Polar Regions Department) were concerned about the very act of strategising in public about the Arctic, and in particular, how such activity might be read by the Arctic states. As I argued in Chap. 3, the circumpolarisation of the Arctic after the Cold War hardened the attitude of the eight Arctic states towards the rest of the international community especially on matters relating to jurisdiction and governance.[7] The Ilulissat Declaration—which affirmed the primacy of the Arctic littoral states in large areas of the Arctic Ocean—was announced only two months after the Oban workshop. Also, in 2008, following a surge in applications for Observer status at the Arctic Council from countries such as China, Japan, and South Korea, the Arctic states announced a review of the contributions of existing Observer states, including Britain, effectively reminding the international community that, in de facto terms, the Arctic was first and foremost the preserve of the Arctic states. Given those geopolitical anxieties, the Government's decision to avoid openly strategising about its interests in the Arctic suggests that it may have found that the potential benefits of producing a formal strategy were insufficient for running the risk of being excluded from one of the most important international bodies for the region.[8]

However, beyond the British Government, others (including, at the time, this author) were not so restrained. Between 2010 and 2013, several calls were heard at workshops and conferences (such as the Canada–UK Colloquium in 2010) that the British Government needed to do more to show that it was taking what was happening in the Arctic, in terms of both risks and opportunities to Britain, seriously. Allusions to strategic thinking occurring behind closed doors were not enough. Stakeholders—ranging from businesses to environmental non-governmental organisations (NGOs) (with high levels of public support)—wanted to see more evidence that the Government was committed to helping them further their interests in the Arctic, and it seemed that only a formal strategy paper would do.

Why a strategy? Because, as I argued together with Klaus Dodds in 2011, the very act of strategising would force the British Government and other stakeholders to reflect on what mattered to Britain in the Arctic, and how those interests might best be realised in spite of ongoing uncertainties about how the Arctic's future might unfold. Moreover, the time-consuming and challenging nature of strategising would itself entail—and, perhaps more importantly, signal to others—a commitment to the Arctic that in turn would reassure potential partners (both domestically and internationally) that Britain was investing in its relations with the Arctic (Depledge and Dodds 2011).

The debate about whether Britain should prepare a formal Arctic strategy or, at the very least, a detailed policy paper climaxed in 2012, when the House of Commons Environmental Audit Committee, spurred on by environmentalist organisations such as Greenpeace, the World Wide Fund for Nature (WWF), and E3G, launched its own inquiry into Britain's role in the

Arctic, and how the British Government might leverage greater influence, specifically in support of its broader international commitments protecting the environment, and encouraging sustainable development. Several submissions of written evidence cited by the Committee in its final report made the case for Britain to have its own Arctic strategy, and the Committee itself concluded that the Government should 'begin the development of an Arctic Strategy to bring together the UK's diverse interests in the Arctic and engage all stakeholders' (Environmental Audit Committee 2012: 68). The Committee also set out in detail what it thought such a strategy should like, for example, by emphasising the necessity of identifying

> potential end-states for the Arctic and how the Government intend (sic) to use its influence at the UN and Arctic Council to bring those about, taking account of the limits on the UK's ability to directly drive such changes. (Ibid.: 68)

Up to that point, successive British governments had remained reluctant to produce any kind of formal policy paper on British interests in the Arctic, despite the fact that, in addition to the 2008 Oban workshop, the Foreign Office was already trying to think more strategically about Britain's interests in the Arctic. That much was evident in an internal report prepared by the Foreign Office Strategy Unit titled 'The Arctic: Strategic Issues for the UK'. But even that was an inherently conservative paper that reported little need to actually address those issues strategically, at least in the short term, beyond improving coordination across Whitehall departments for the purpose of monitoring developments (Foreign Office 2009). The Polar Regions Department did publish a 900-word statement on 'The UK's Engagement in the Arctic' on its website in 2011, and made tentative steps towards producing a more detailed policy paper by summarising British interests in briefing papers produced for the Foundation for Science & Technology and the Environmental Audit Committee, but ministers and officials continued to reject the idea of a publishing a formal strategy.

Getting Back on the Front Foot

In Britain, it is within the power of parliamentary Select Committees to demand that the British Government responds to conclusions and recommendations arising from their inquiries. Consequently, following the publication of the Environment Audit Committee's report on the Arctic in September 2012, the Government had to give a response to the Committee's call for Britain to develop its own Arctic strategy. Now under pressure from Parliament, the Government agreed in 2012 to change its approach in the Arctic, although it was still reluctant to talk about strategies.

Was it all down to Parliamentary pressure that the British Government changed its approach? It seems unlikely, although it may well have been the final straw. After all, the Polar Regions Department had already been making ever-longer statements about British interests in the Arctic, suggesting, perhaps, that it was heading in the direction of producing a more formalised policy paper anyway. In June 2013, shortly before the Government reported back to the Environmental Audit Committee, I learned that the decision to move to a formal policy paper had been embraced for at least two reasons: the first reason was that the inquiry had raised the profile of the Arctic across Government as it forced relevant departments and agencies to engage with issues such as energy, shipping, and environmental protection. That in turn would make it easier for the Polar Regions Department (which was charged with the responsibility for coordinating and drafting the paper) to gather cross-Whitehall input. The second, and perhaps more significant reason, was that the Environmental Audit Commitment had created a useful opportunity for the Government to 'get back on the front foot' with its Arctic messaging.[9] The desire, or perhaps even the *need*, to do so suggests that the Polar Regions Department, which was nominally responsible for communicating Britain's interests in the Arctic, had become increasingly frustrated by its apparent inability to shape how Britain's role interests in the Arctic were being defined and represented.

That British Government officials were doubting their ability to dominate how Britain's role and interests in the Arctic were being defined and represented was a testament to the strength of environmental organisations that had been campaigning for decades to protect the Arctic's environment from the polluting effects of human activity. Environmental NGOs have been prominent in Arctic geopolitics for decades. WWF, for example, have operated an Arctic programme since 1992 with offices in six Arctic countries and several Observer states (including Britain). It also has Observer status at the Arctic Council. Greenpeace UK has an even longer history in the region having first campaigned against whaling in Norwegian and Icelandic waters in the late 1970s. Despite a shift of focus to Antarctica in the 1980s and 1990s, over the past decade it has become increasingly active in trying to encourage successive British governments to take a stand against oil and gas operators seeking to exploit Arctic reserves, as well as trying to get the fishing industry to act more responsibly in Arctic waters.

In August 2010, four Greenpeace UK campaigners managed to shut down an oil rig operated by Cairn Energy, a British-based oil exploration company, off the coast of Greenland. That happened at the beginning of a more extensive campaign against British-based oil companies with assets in the Arctic, most notably Shell. In 2012, following the United Nations (UN) Rio+20 Summit, Greenpeace International announced it was moving to a 'war footing' to counter efforts to open up the Arctic to offshore oil drilling and unsustainable fishing (Black 2012). That included launching the 'Save the Arctic' campaign, which, with massive celebrity endorsement, sought a million signatures on a petition calling for greater protection of the Arctic. Greenpeace also stepped up its direct action targeting Shell's operations in Alaska, and Gazprom's *Prirazlomnoya* oil rig in the Russian sector of the Barents Sea, while also carrying out other dramatic stunts in London.

Such activism put the British Government in a difficult position for several reasons. The first is that several campaigns against Arctic oil and gas have been started, or at the very least attracted significant attention from the media and the wider public, in Britain. The second is that two of the world's biggest oil majors, BP and Shell, have headquarters in London, while the smaller operator, Cairn Energy, has its headquarters in Edinburgh. All three companies had been actively investing in Arctic exploration as a consequence of growing interest in the region's oil and gas resources. The past decade has seen numerous demonstrations across Britain against these firms. The third is that the 2010 Deepwater Horizon explosion and oil spill was not just damaging to BP's reputation, but also to the British Government's. That was in part because then-President of the United States, Barack Obama, insisted on referring to BP by its former name 'British Petroleum', and in part because, just a few months after that disaster, the Government had publicly supported a new 'Global and Arctic Strategic Alliance' between BP and the Russian state oil company Rosneft—a move which was savaged by Platform, another British-based environmentalist pressure group, and Greenpeace (Platform 2011). The fourth is that throughout these events, and despite the Government's commitments to reduce British greenhouse gas emissions and lead international efforts to tackle global climate change, the Department of Energy and Climate Change insisted that Britain could not rule out support for

Arctic oil and gas development as such activity might prove crucial for realising its broader energy goals of diversifying global energy supplies and maintaining stable energy prices.

With their attention-grabbing stunts and celebrity endorsements, the environmental groups had moved to the forefront of British public debate about the Arctic and what Britain's role there should be. Yet in their calls for an international 'Arctic sanctuary', environmentalist groups were on difficult ground. The A8 were hostile to any attempt to devise a global instrument to govern even those areas of the Arctic that are considered part of the global commons. There was, then, perhaps a perception in Government, that Britain, by indirect association with the noises coming from environmental groups, risked being seen as a problematic partner in the Arctic, especially since it still had not spelled out its position on key Arctic issues relating to climate change and sustainable development.

That risk only grew when the Environmental Audit Committee—which, as already mentioned, was working closely with Greenpeace and other environmentalist organisations—announced that its inquiry would focus on what more Britain could do to 'protect the Arctic', effectively implying that the region might be in need of British protection if the Arctic states were found wanting. The final report would have done little to dispel such concerns either as the Committee called on the British Government to demand a moratorium on drilling in the Arctic until certain conditions were met—conditions that were devised to satisfy the Select Committee of a parliamentary chamber in a country with no sovereign jurisdiction in the Arctic.

For the British Government, then, the statutory requirement to respond to the Environmental Audit Committee's call for a 'strategy' presented a timely opportunity to wrest the narrative of Britain's role in the Arctic from environmentalists. Specifically, the Polar Regions Department wanted to communicate a more balanced view of Britain's interests in the Arctic—one that encompassed British interests in science and commerce, as well as concerns about environmental protection and sustainable development. The subsequent decision to publish an Arctic 'policy framework' (but note, not a 'strategy'; see below), while motivated by the Committee's inquiry, was therefore not along the lines that the Committee envisaged.

That much was evident in the heated exchanges that occurred in July 2013 when Mark Simmonds (Permanent Under-Secretary of State for Foreign and Commonwealth Affairs, 2012–2014) appeared before the Committee to defend the British Government's response. The dispute centred precisely on what kind of role Britain should be playing in the Arctic. The Committee had refused to accept that environmental protection and oil and gas development in the Arctic were compatible goals, and that, consequently, Britain's primary role in the Arctic should be to push firmly for tougher environmental measures that included promoting a moratorium on further oil and gas exploration. The Committee's position also reflected the stance of the various environmental organisations that the Committee had been working with. However, the Government remained unswayed in its view that Britain could play a balanced role, supporting measures to promote both environmental protection and oil and gas development, in the Arctic—what appeared to be at issue for the Government though was the need to communicate that position, and the judgements on which it was formed, with greater transparency and effectiveness.

The Arctic Policy Framework that was published in October 2013 was therefore not a revolutionary document. As the British Government communicated in its response to the Committee, the Arctic Policy Framework was to be 'a summary of the policy framework that we have in place at the moment', one that would 'pull together all the different strands' of British interest in the Arctic, 'demonstrate the significance and importance we attach to [Arctic issues]', 'communicate [the Government's] Arctic policy effectively', and make 'more accessible [Britain's] Arctic policies'. Its purpose would be to explicate the Government's existing relationship with the Arctic, rather than to radically redefine it, as the Committee was hoping for. That meant reaffirming the Government's policy of taking what it saw as a balanced and pragmatic approach to a range of issues in the Arctic that reflected Britain's environmental, scientific, and socio-economic interests in the region, while, owing to Britain's status as a 'non-Arctic' state, remaining mindful of the limits of British influence.

The policy framework consequently couches British interest in the Arctic carefully. Concerns about the environment are balanced with opportunities associated with the possibilities of socio-economic development in the Arctic. Repeated references are made to Britain's respect for the sovereign rights and interests of Arctic states and Arctic peoples, as if that respect had somehow come to be doubted as a consequence of public and parliamentary support for an environmental sanctuary in the Arctic. Where it seeks to legitimise British interests, it does so on the basis of both Britain's topographical proximity to the Arctic and far-reaching 'Global Arctic' topologies that connect events in the Arctic to broader processes of environmental change, commercial activity, and international legal regimes in which Britain plays an active role. The vision that the British Government proffers of a 'safe and stable', 'well-governed' Arctic, where 'policies are developed on the basis of sound science', and only 'responsible development takes place', is one that the Arctic states could have written themselves, and conjures up a future, which, on the face of it, few, if any, would ever be likely to contest.

What's in a Name?

The British Government's decision to publish a 'policy framework' instead of a 'strategy', as called for by the Environmental Audit Committee and several others, was a deliberate one (see also Depledge 2013). Yet, at the same time, the Government has played on the vagueness of the term 'policy framework' to communicate different messages to different audiences. That has produced confusion among stakeholders over what the policy framework actually is and what it is trying to achieve. After all, Mark Simmonds implied in the evidence he gave to the Environmental Audit Committee in July 2013 that while the policy framework was to all intents and purposes a strategy, the Government had to be 'careful with the language'. The Polar Regions Department had similarly told the Committee at an earlier evidence session that they were 'slightly playing semantics' by avoiding the term strategy yet that did not mean the Government did not have a 'clear Arctic policy', or a clear view on what it 'want[s] to get out of the Arctic' (e.g. scientific knowledge and commercial gain). Those statements are demonstrative of how the Government has responded in general to domestic stakeholder demands for action by claiming that the policy framework is effectively a strategy.

However, such statements are also indicative of the British Government's continued reluctance to talking openly about strategising in the Arctic. That reluctance appears to stem at least in part from successive British governments' concerns about the sensitivities of certain Arctic states to any act by a non-Arctic state that might explicitly or implicitly contest the primacy of the A8 in Arctic affairs (the origins of those sensitivities were discussed in Chap. 3). These concerns were laid bare in the evidence that the Polar Regions Department gave to the House of Lords Arctic Select Committee in 2014. When asked about the origins of the policy framework, Jane Rumble (Director, Polar Regions Department, 2007–ongoing) explained:

> Some of the Arctic states felt that a 'strategy' is connected to something over which you have direct control and has objectives and deliverables, and felt that it might be going a bit far for a non-Arctic state to suggest that it was in control of various elements of Arctic policy. Some of the other Arctic states were not quite so sensitive, but to walk that particular tightrope, we decided not to call it a strategy but to set out our Arctic policy interests in a framework. (Arctic Committee 2015: 244)

And yet, despite what British Government ministers and officials have said, the policy framework is clearly not a strategy, either in terms of the curious definition of a strategy apparently offered by some of the Arctic states (as relating to direct control of elements of Arctic policy) or in the broader sense of the 'art of creating power' as outlined at the beginning of this chapter. Instead, as noted earlier, the policy framework is an explication of Britain's existing Arctic interests and policies, framed mostly in terms of continuity rather than change. While the document recognises that uncertainty, instability, and unpredictability will shape the future of the Arctic, there is little in it that suggests the Government intends to use that to advance Britain's interests. Instead, the emphasis is on supporting the Arctic states in maintaining the geopolitical status quo, and on being ready to respond to any changes. Challenging that status quo might of course risk a backlash from the Arctic states but failing to challenge it suggests that the Government is more concerned about maintaining amicable relations with the Arctic states than it is about trying to take the initiative in shaping future outcomes in the Arctic that enhance, rather than simply conserve, Britain's position in the region. The point here though is not to pass judgement on the Government for taking a conservative position.[10] Rather, it is to make clear that the Arctic Policy Framework is not a strategy, and that the Government's attempt to fudge the semantics seems to be aimed more at placating domestic audiences—which demand, increasingly, 'strategies' for every challenge facing Government[11]—than at disguising strategic content.

A Strategic Act Nonetheless?

The Arctic Policy Framework may not be a strategy, but that does not mean it is without strategic effect—or indeed strategic *affect*—as it sought to reassure the Arctic states about the nature of Britain's interests. And just because a formal strategy paper has not been published does not necessarily mean that strategic thinking about the Arctic does not go on behind closed doors. In fact, the production of the Arctic Policy Framework seems to be, at least partly, a performative act by the Polar Regions Department to try to work around certain challenges that have frustrated their attempts to communicate Britain's Arctic interests both domestically and internationally.

The first of those challenges stems from the way in which, across the British Government in general, emerging opportunities and risks in the Arctic have continued to be treated as distant concerns. Alex Evans, a member of the British diplomatic service and a visiting research fellow at King's College London, has highlighted how 'short termism' and 'recentism'—which he describes as a focus on the immediate past at the expense of a longer view—continue to dominate British foreign policy (Evans 2014). He goes on to argue that, by focusing too much on the present and the recent past, the Government has reduced the time and resources available for thinking strategically about the future. That in turn produces behavioural biases that favour policies which maintain or reinforce the status quo. The Norwegian political scientist and social anthropologist Iver Neumann has explored similar themes in his ethnographic study of the foreign policy establishment in Norway where the need for internal consensus invariably relegated innovative moves in favour of preserving existing, and reinforcing, policies and practices (Neumann 2007).

Such conservatism is inherent in the Arctic Policy Framework. To some extent, the viability of actually producing an Arctic Policy Framework depended on the Polar Regions Department's ability to draft a paper that the whole of government could sign up to. The process involved consultations between Polar Regions Department officials and all other British Government departments with policy responsibilities that have, or could have, an Arctic dimension.[12] That was no easy task and it was up to the Polar Regions Department to press other government departments to take ownership of those sections of the paper that were relevant to their policies. One official told me that the process would likely have been much harder had the Environmental Audit Committee not called on the Government to publish an Arctic strategy because it helped the Polar Regions Department pressure other government departments into engaging.[13] However, given the mixed level of interest across Government, the Polar Regions Department also had to be careful not to ask too much of other government departments, for example, by asking for detailed policies about a subject that those departments did not deem to be of pressing concern.

There was consequently very little in the document to suggest that Britain was going to change its approach to the Arctic in response to recent developments and ongoing uncertainty about the region's future. While the Arctic Policy Framework offered several examples of what Britain was already doing in the Arctic in terms of science and business, it was not always clear about what means those ends are connected to. Moreover, while there were various statements about what Britain will (or wants to) do in the Arctic, there was very little information about how it would do those things with the means to hand. In general, the Arctic Policy Framework was largely non-committal and made few material demands for greater resourcing of Britain's Arctic policies across the various departments and agencies involved.

What the Arctic Policy Framework left us with then was a document that was so lacking in depth of detail that it gave the British Government (and individual departments and agencies) considerable wriggle room in terms of how, and indeed whether, Arctic interests were pursued. Others might argue that makes the Arctic Policy Framework adaptable—and that certainly seemed to be what Mark Simmonds was getting at when he described the Arctic Policy Framework as a 'living document' that would be regularly reviewed as conditions in the Arctic changed and Britain's interests in the region evolved. However, at the time of writing, the first review of the Arctic Policy Framework since 2013 has still not been published despite major changes affecting the region including record sea ice loss, economic sanctions on Russia, the election of Donald Trump, and a dramatic decline in the global oil price. Meanwhile, the detail was effectively kicked into the long grass, to be left there until Arctic issues became more pressing than the Government's other priorities in traditional arenas of British foreign policy. The issue of needing to resource Britain's Arctic policies was thus also put into abeyance.

The Arctic Policy Framework therefore imposed very little cost on the British Government, and that in turn helped make it a white paper that all departments could get behind. At the same time, from the strategic perspective of officials in the Polar Regions Department, they now had a document—which they have at times implied is a 'strategy'—that they could use to show their detractors domestically that Britain was committed to its relationship with the Arctic, while at the same time clarifying the Government's position on Britain's interests in the region to both domestic and international audiences.

The second strategic challenge that the Polar Regions Department sought to overcome by producing the Arctic Policy Framework was the growth of international interest in the Arctic, and the threat which that posed to Britain's continued influence in the region. The timing of the Arctic Policy Framework's publication assumed even greater significance as it came just months after an Arctic Council Ministerial meeting in Kiruna in May 2013, where the A8 agreed to admit six new countries as Observers. Britain had long regarded itself as the 'lead Observer' in the Arctic Council, following its scientific contributions, but that position became increasingly precarious as the Arctic's international profile grew. Increasing interest in the region, especially from Asian states such as China, Japan, Singapore, and South Korea, led to a growing sense that Britain and other European countries that had perhaps taken for granted their involvement in Arctic affairs risked having their interests crowded out by new players. At the same time, the Arctic states were tightening their grip on the Arctic, for example, with the introduction of new measures designed to ensure that Observer states at the Arctic Council showed due deference to the primacy of their sovereignty and interests in the region. As a consequence, the British Government, and particularly the Polar Regions Department, had to work much harder to explain and justify Britain's role and interests in Arctic affairs.

The act of producing the Arctic Policy Framework appears to have tried to change that situation in two ways. The first was to frame Britain's role in the Arctic as being that of a 'model neighbour/Observer state' (Depledge and Dodds 2014). That was apparent in the way that British Government officials have been careful about the use of the term 'strategy', as well as in the way the document itself explicitly states Britain's deference to the geopolitical status quo in the Arctic. Those moves helped the Government avoid provoking those Arctic states (notably Canada and Russia) that have been most anxious about how increased interest from beyond the Arctic might affect their sovereignty and authority. They also stood in stark contrast to the debacle surrounding the EU's flawed attempts to draw up a strategy for the Arctic, and the suspicions that Chinese interest in the Arctic had provoked owing to a perceived lack of transparency about its intentions in the region, further helping to enhance Britain's reputation among the Arctic states (the Arctic Policy Framework was reportedly received well by the Arctic states).[14]

In tandem with the move to position Britain as a 'model' Arctic neighbour, the Arctic Policy Framework was also used by the British Government to remind the Arctic states that Britain was, topographically speaking, the Arctic's 'nearest neighbour'. As more and more states, from Asia (China, India, Japan, South Korea, and Singapore) and Europe (France, Germany, Italy, and Poland, among others) have demonstrated greater interest in the Arctic (through bilateral agreements, applications for Observer status at the Arctic Council, and material investments in science and infrastructure in the Arctic), there appears to have been a greater effort from the Government to distinguish Britain's Arctic credentials. In lieu of financial and political capital, the 'nearest neighbour' terminology worked to naturalise Britain's 'near-Arctic' identity in a way which girds the legitimacy of British interest(s) in the Arctic, while at the same time distinguishing its interest from that which is emerging in other parts of the world.

However, since the Arctic Policy Framework was published, there have been four further Select Committee inquiries to scrutinise whether the British Government's approach was the right one. Already, exchanges heard during the inquiries by the House of Lords Arctic Select Committee, Environmental Audit Committee, Scottish Affairs Committee, and Defence Committee show that Parliament still has divergent ideas about what kind of place the Arctic is, and what kind of role and interests Britain has there. The House of Lords Arctic Committee Select Committee inquiry perhaps came too soon after the publication of the Arctic Policy Framework to be able to challenge the Government's approach, while the other three inquiries were shelved in May 2017 because a General Election was called. Even though it remains to be seen if they will be relaunched, the very fact that there seems to be so much interest in Parliament in scrutinising the Government's approach to the Arctic implies that this in an area where the Government will likely be pressed to do more in the coming years. Of course, what that 'more' looks like is still to be determined as actors from both within and beyond Parliament continue to produce overlapping and conflicting ideas about what matters most to Britain (and, an increasingly independently minded Scotland) in the Arctic.

Conclusion

The British Government remains in a quandary over how best to define, represent, and ultimately pursue its Arctic interests. Calls from Parliament, environmental NGOs, academics, businesses, and other stakeholders for the Government to devise an Arctic strategy seemingly reflect a desire to see Britain 'do something' about the challenges and opportunities emerging in the Arctic, and ensure that the Government is held accountable for its actions (or inaction) towards the region. Although there is still uncertainty over what the region's short-term future will be, such calls perhaps reflect a sense of anxiety that Britain is losing influence in the Arctic—and the world—or, conversely, a sense of entitlement to a role in Arctic affairs owing to a combination of history, geography, and contemporary power (both of which have been evident in abundance during the fractious debate about Britain's future place in the world after it is due to leave the EU). But both lines of reasoning point to the same concern: Britain is at risk of missing out altogether on opportunities to shape the Arctic's future.

The Polar Regions Department, as the subsection of the Foreign Office effectively responsible for representing Britain in Arctic affairs, has borne the brunt of such criticism, despite having only limited resources with which to execute any change of approach. After all, the Department's primary role is still to look after Britain's Antarctic interests. While it does act as the coordinator of a cross-Whitehall network of departments and agencies, its ability to effect change depends on whether those other parts of Government are willing to invest their own resources into supporting British policies in the Arctic. That is something which both 'short termism' and 'recentism' will continue to work against unless there is another major development in Arctic affairs, such as a first completely ice-free summer in the region. Furthermore, an apparent lack of Cabinet-level interest in British Arctic policy only makes it harder to resist pressure from the A8 that Britain, along with everyone else, should show deference to the geopolitical status quo in the Arctic, including the primacy of their sovereignty and interests in the region.

The Arctic Policy Framework should be regarded as the Polar Regions Department's first substantial attempt to navigate these pressures and communicate a clear statement of Britain's role and interests in the Arctic. It sought to do so in at least three ways. The first was to unify the British Government's approach to the Arctic in a single document. It is significant that the final document was published by HM Government, and not the Foreign Office, because it showed that the whole of Government had signed off on it. The second was to get the Government back on the front foot with its Arctic messaging. It now had a white paper that it could use to show detractors that Britain was committed to pursuing its Arctic interests, while also creating a new reference point for anyone wanting to learn more about British interests in the region (prior to the paper, that reference point would likely have come from non-government sources, i.e. academics or civil society). The third was to communicate to the Arctic states that Britain was a 'model neighbour' and that the Government was in no way encroaching on the sovereignty and interests of the Arctic states by setting out its interests in the region. Thus, although the Arctic Policy Framework was not a strategy, it was still a strategic act, aimed at restoring the Government's primacy in defining how Britain's role and interests in the Arctic are constructed, represented, and narrated to both domestic and international audiences.

But the Arctic Policy Framework remains a precarious framework through which to define Britain's role and interests in the Arctic, and could even prove to be a hindrance to the British Government in the longer term. Challenges to the Arctic Policy Framework are already arriving from several directions. Since it was published in 2013, there have been several developments in the Arctic with implications for British interests in the region. That includes the dramatic decline of the global oil price, which has put several oil and gas development projects in doubt, especially in the North American Arctic where Shell was active until 2015. It also includes the deterioration of relations between Russia and the United States and the EU. Economic sanctions imposed by the United States and EU, supported by Britain, have specifically targeted Russia's ability to develop oil and gas projects in the Arctic, affecting British companies such as BP that have close partnerships with Russian firms. Meanwhile, heightened Russian military activity in the Arctic is causing concern at a time when the Armed Forces remain vulnerable to cuts that affect their ability to defend Britain's Northern Flank. The Arctic has also experienced further record-breaking temperatures and sea ice decline. The Government, though, is yet to acknowledge the implications of these developments by revising the Arctic Policy Framework, despite the commitment the Foreign Office made to the Environmental Audit Committee in 2013 that it would be a responsive document. Although a revised Arctic Policy Framework is due to be published at the end of 2017, four years after the first iteration, in general it has proven to be no more 'living' than other Government papers that get revised on similar timescales, such as the National Security Strategy.

Lastly, in its current form, the Arctic Policy Framework arguably shows too much deference to the geopolitical status quo in the Arctic, and the interests of the Arctic states. The Polar Regions Department might hardly have been expected to do otherwise, given its limited resources, and the apparent lack of high-level interest in the Arctic (i.e. from the Foreign

Secretary, the Cabinet Office, or the Prime Minister's Office). Regardless though, the effect of positioning Britain as a model neighbour and accepting implicitly that an alternative approach might be considered threatening to the interests of the A8 is that it reinforces the processes of circumpolarisation that have been under way for the past decade or so. By agreeing not to openly strategise about the Arctic, the British Government effectively legitimises the view held by some Arctic states that the interests of other countries from beyond the region should be subordinate to those of the Arctic states, even in those areas that lie beyond the sovereign jurisdiction of the A8. In establishing that precedent, it will be much harder for Britain to contest the circumpolarisation of the Arctic in the future, should that be in its national interest to do so. There is then a lack of foresight in the Arctic Policy Framework about the permanency of British interests in the Arctic, despite there still being great uncertainty over what the Arctic's future will be. As geopolitical and material conditions in the Arctic continue to evolve, as they inevitably will, the Arctic Policy Framework, in its current form, will only constrain Britain's ability to adapt.

Notes

1. The other questions posed were 'Is the UK collectively maximising its interests and opportunities in the Arctic?', 'What are the future changes and challenges in the Arctic? How should the UK respond to these?', and 'How do we ensure that we maximise the benefits arising from UK science in the region?' (GBSC 2008: 3).
2. Today, changes in the Arctic are of interest to the Foreign and Commonwealth Office (diplomacy), the Ministry of Defence (defence and security), the Department for International Trade (trade), the Department of Business, Energy & Industrial Strategy (business, resources), the Department for Environment, Food & Rural Affairs (fisheries, biodiversity, conservation), and related agencies and advisory bodies. The devolved Scottish Government has also expressed its interests in the region.
3. Budget cuts and a blurring of domestic and international responsibilities across Government, over recent decades, has resulted in many departments and agencies adopting international profiles of their own (and subsequently taking the lead at international negotiations). Britain is also represented by the EU in some areas, such as fisheries, and will continue to be until it formally leaves.
4. Author interview with Foreign Office official, 5 October 2011.
5. Author interview with Foreign Office official, 5 October 2011.

6. After all, the British Government does not have strategies for other regions like 'Asia', 'Africa', 'North America', 'Latin America', or the 'Middle East'.
7. Not all of the Arctic states were resistant to the international community becoming more engaged in Arctic affairs, but the importance of reaching a circumpolar consensus on the matter meant assuaging what were primarily Russian and Canadian concerns.
8. Author interview with Foreign Office official, 5 October 2011.
9. Author interview with Foreign Office official, 20 June 2013.
10. Broader commentary about the British Government's ability to think strategically suggests that the Government has generally shown itself to be more concerned about managing risks than seizing opportunities when it comes to uncertain futures (Edmunds 2014).
11. Yet as critics of 'over-selling' strategy might argue, what might really be being demanded is a technocratically derived, linear solution to a complex, contingent, and evolving issue, rather than actual strategic thinking (Edmunds 2014).
12. While the Arctic may only enjoy low priority across Government, its pervasiveness is evident in the fact that changes in the region are relevant to several government departments, agencies, and advisory bodies including the Foreign and Commonwealth Office, the Ministry of Defence, the Department of Environment Food and Rural Affairs, the Department of International Trade, the Department for Transport, the Department of Business, Energy and Industrial Strategy, the Maritime & Coastguard Agency, and the Joint Nature Conservation Committee.
13. Author interview with Foreign Office official, 20 June 2013.
14. It is worth noting that following the publication of the Arctic Policy Framework, several other non-Arctic states including Germany, Japan, and the Netherlands have published similar papers.

References

Arctic Committee. 2015. *Select Committee on the Arctic: Oral and Written Evidence*. Accessed August 1, 2017. http://www.parliament.uk/documents/lords-committees/arctic/responding-to-a-changing-arctic-evidence.pdf

Black, Richard. 2012. Rio+20: Sir Paul Backs Greenpeace Arctic Campaign. *BBC News*, June 21. Accessed June 22, 2017. http://www.bbc.co.uk/news/science-environment-18531697

Depledge, Duncan. 2013. What's in a Name? A UK Arctic Policy Framework for 2013. *The Geographical Journal* 179: 369–372.

Depledge, Duncan, and Klaus Dodds. 2011. The UK and the Arctic: The Strategic Gap. *The RUSI Journal* 156: 72–79.

———. 2014. No "Strategy" Please, We're British: The UK and the Arctic Policy Framework. *The RUSI Journal* 159: 24–31.

Edmunds, Timothy. 2014. Complexity, Strategy and the National Interest. *International Affairs* 90: 525–539.

Environmental Audit Committee. 2012. *Protecting the Arctic*. London: The Stationary Office Limited.

Evans, Alexander. 2014. Organising for British National Strategy. *International Affairs* 90: 509–524.

Foreign Office. 2009. *The Arctic: Strategic Issues for the UK*. London: Foreign and Commonwealth Office.

Freedman, Lawrence. 2013. *Strategy: A History*. Oxford: Oxford University Press.

FST. 2011. *Knowledge into Action; Development in the Arctic Region*. Foundation for Science and Technology. Accessed October 27, 2016. http://www.foundation.org.uk/events/pdf/20111214_summary.pdf

GBSC. 2008. UK-Arctic Stakeholders Report of Conference Held at the Scottish Association for Marine Sciences, Oban, March 10–12. Joint Nature Conservation Committee. Accessed October 27, 2016. http://jncc.defra.gov.uk/pdf/gbsc_0809arcticbiodiversityobanreport.pdf

HM Government. 2013. *Adapting to Change: UK Policy Towards the Arctic*. London: Foreign and Commonwealth Office.

Käpylä, Juha, and Harri Mikkola. 2015. *On Arctic Exceptionalism*. Helsinki: The Finnish Institute of International Affairs.

Neumann, Iver. 2007. "A Speech That the Entire Ministry May Stand for," or: Why Diplomats Never Produce Anything New. *International Political Sociology* 1: 183–200.

NERC. 2007. *Next Generation Science for Planet Earth: NERC Strategy 2007–2012*. NERC: Swindon.

Platform. 2011. *Arctic Anxiety: BP, British Foreign Policy and the Rush for Polar Oil*. London: Platform.

CHAPTER 6

Conclusions

Abstract This concluding chapter offers a brief summary of the main arguments made in earlier chapters about Britain's stake in the Arctic. It then offers some final thoughts on the changing context and challenges facing Britain as it seeks to find a lasting role for itself in the region. Specifically, the chapter argues that Britain needs to invest more in increasing its topological proximity to the Arctic if it is to avoid being pushed to the periphery of Arctic affairs. The chapter also considers how growing usage of the terms 'Global Britain' and 'Global Arctic' might be used to draw more attention to British interests in the region. Lastly, it emphasises the need to situate Britain's search for a role in the Arctic within a progressive sense of what the Arctic might become, rather than what it once was.

Keywords Global Britain • Global Arctic • Peripheralisation • Proximity • Progressive • Investment

More than 1000 years have passed since Orthere of Hålogaland visited the court of King Alfred the Great of Wessex to trade on his knowledge of the Arctic. But Alfred chose to distance himself and his kingdom from those northerly lands and seas, leaving it to others to shape the emerging geopolitical land-, ice-, and ocean-scapes of the far north. Yet in the centuries that followed, successive rulers and governments of the British Isles brought the Arctic closer to home, folding together British imperial, commercial,

© The Author(s) 2018

D. Depledge, *Britain and the Arctic*,
https://doi.org/10.1007/978-3-319-69293-7_6

religious, and scientific interests with Arctic sites, inhabitants, resources, and passageways. Over that period, the Arctic came to function as several different things to Britain: an abomination, a transit route, a scientific laboratory, a resource frontier, a military theatre, a vulnerable environment, a home to indigenous people, and a place for adventure. Britain and the Arctic came to share a substantial history, forged not by physical geography, but by a shifting web of connections that brought the Arctic into the very heart of towns and cities across Britain, and left Britain etched into both imaginary (i.e. in the scientific discoveries and on the maps that continue to bear labels ascribed by British explorers) and physical land-, ice-, and ocean-scapes (i.e. the wrecks of ships and the damage caused by industrial pollution) of the Arctic.

Today, these memories of Britain as an Arctic nation seem to have faded from view, at least for many of those working in foreign policy and international affairs whose priorities appear to be elsewhere. Those few memories that do persist are often found in novels, films, and television programmes—what some might describe as a banal Arctic nationalism that is more subconscious than conscious. British polar scientists and explorers can still be awarded the 'Polar Medal' (originally known as the 'Arctic Medal'), connecting modern-day feats of endurance and science to the legacies of nineteenth-century explorers. The Scott Polar Research Institute in Cambridge also stands as a permanent reminder of Britain's polar heritage to both the Institute's staff and the wider public. Elsewhere, specialist historians, librarians, and museum curators continue to tell stories about intrepid explorers and the monstrousness of the Arctic that they encountered on their fateful journeys north. Such stories are still expected to shock and appal their audiences as they are invited to consider the hardship and the desperation that explorers must have faced, but not necessarily the geopolitical imperatives that drove them to the Arctic in the first place, or why the region continues to matter to Britain today.[1]

Even reading a news story about the Arctic today is frequently to read a story of personal or collective accomplishment as individuals test their ability to survive and thrive in one of the world's harshest environments. Implicit within such stories is the idea of an Arctic hostile to human life. But at the same time, the Arctic also seems to be regarded as a much more benign space now, as luxury cruise ships slip through the Northwest Passage with ease, and television channels broadcast footage of the Arctic's flora and fauna. Increasingly, though, the idea of the Arctic as a benign space has also tipped into something else: concern about the Arctic's vulnerability as the region is overcome by the climatic impacts of what is being widely termed the 'Anthropocene'—the notion that human beings are now the most decisive influence on what happens in the natural world. Over the past decade or so, the Arctic has often been described as the canary in the coal mine of climate change.

This book has sought to balance these different understandings of the Arctic as a testing, benign, or vulnerable geographic region with an account of how the Arctic is also still being rendered important to Britain in geopolitical terms—to start a more mature conversation about why the Arctic matters to Britain today. Instead of continuing to feed the public imagination with stories about lost explorers and their artefacts, which continue to showcase and foreground the legacies of the British Empire, my focus is on Britain's contemporary interests. Specifically, I have argued that Britain, which was once a dominant presence in the Arctic by virtue of its colonial possessions, expansive trade networks, strategic interests, and expeditionary activities, has over the last century or so been pushed to the periphery of Arctic affairs (Chap. 2). That has happened because of a general ambivalence about the Arctic relative to Britain's other foreign policy priorities, and because Arctic affairs have increasingly been defined by the principle of circumpolarity as the self-styled A8 have sought to firm up their primacy in the region (Chap. 3). Yet Britain still has significant commercial, scientific, and military interests in the Arctic that carry with them the potential to connect Britain to the Arctic in new ways that stretch and test the limits of circumpolar definitions of the Arctic, opening up the possibility of a greater role for Britain (and others) in the Arctic in the future (Chap. 4). Seizing that opportunity will require a sustained political and financial commitment to the region that is mindful of the sensitivities of the A8, without being overly deferential to them (Chap. 5).

Instead of further summarising earlier material presented in this book, what follows next are some concluding thoughts on the changing context and challenges facing Britain as it seeks to find a lasting role for itself in the Arctic in the twenty-first century.

Being Near-Arctic

Throughout this book I have used the concept of 'proximity' to explore what it is that brings Britain and the Arctic together in the early twenty-first century, how these connections have evolved over time, and the barriers that have emerged. Specifically, proximity has been defined in two ways. The notion of topographical proximity has been used to refer to the physical distances that separate discreet geopolitical containers of space, including 'Britain' and the 'Arctic'. That physical distance is assumed to have become more or less fixed as geopolitical borders have assumed greater permanency in most of the world (although there is always the possibility that Scotland will break away from the rest of the United Kingdom, increasing the topographical distance between Britain and the Arctic). In line with that topographical geography, over the past decade it has become commonplace for British Government officials to refer to Britain as the Arctic's 'nearest neighbour', as if Britain's physical distance from the Arctic relative to other non-Arctic states privileges British interests in the region.

At the same time, the Arctic Policy Framework published in 2013 acknowledged—but arguably underplayed—that British interest in the Arctic is rooted in more than its physical proximity to the region. That is because the Arctic also matters to Britain because of the numerous ways—scientifically, environmentally, commercially, strategically—that the two geopolitical entities are connected. The notion of topological proximity rests on this idea of connectivity and is used to show how two entities can seem closer together, or more distant from each other, because of how connected they are.

Privileging topographical proximity over topological proximity, as happens implicitly whenever someone leads their account of Britain's Arctic interests by declaring that Britain is the Arctic's nearest neighbour is a risky proposition in contemporary Arctic geopolitics. Doing so implies that Britain will always have its topography to fall back on when it wants to justify its interests in the Arctic, and that this will be accepted by the A8. However, the A8 are being cautious about embracing the rest of the international community, and are increasingly asking non-Arctic states to demonstrate their value to the Arctic in return for their presence in organisations like the Arctic Council. This suggests that beyond the geography of the Circumpolar Arctic, topography matters far less than topology as the A8 look to make connections with the rest of the world that best support their Arctic interests, regardless of how far away from the Arctic Circle those connections take them. In contrast, the privileging of topographical proximity downplays the importance of actively building connections to the Arctic for extending Britain's influence in the region, whether that involves funding scientific research, defence engagement, diplomatic visits, or trade missions on the grounds that Britain already shares some sort of naturalised affinity with the Arctic owing to its geography.

The notion that Britain is somehow more proximate to the Arctic on the basis of its topographical geography is further undermined by the fact that, as the British Government itself seems keen to argue, the Arctic is not heterogeneous. It follows that if the Arctic is not a heterogeneous space, then being topographically near to one part of the Arctic does not necessarily make you topographically near to another part. In other words, while Britain might claim to be the nearest neighbour of the 'Atlantic Arctic', that certainly cannot be said of the 'Pacific Arctic', encompassing Alaska and the Russian Far East, which topographically is much closer to countries such as China and Japan. That would again suggest that how invested and how connected a country or any other site is to different parts of the Arctic matters far more than the ability to claim topographical proximity to a relatively small part of the region.

If Britain is to have a bigger impact on Arctic affairs, the British Government should focus less on claiming that Britain is the Arctic's *nearest* neighbour and focus more on making—through active investment in science, defence, trade, and cultural links—Britain the Arctic's *closest* neighbour. As I have observed elsewhere, the word 'making' implies the need to actively construct and cultivate Britain's relationship with the Arctic, rather than assuming it exists a priori (Depledge 2013). At the same time, the conceptual shift from 'nearness' to 'closeness' (two terms that Government officials currently seem to use interchangeably) is more than intellectual gymnastics because it is intended to draw out the difference between two objects being near to each other, and two objects being close to each other. In the case of the former, the objects may be topographically proximate but otherwise unconnected and independent. In the case of the latter, the objects may be topographically distant, yet connected to each other in innumerable ways that make them mutually dependent. Put simply, the more connected that Britain is to the Arctic, and the more it invests in those connections, the closer the Arctic will be. Conversely, the less Britain invests in its connections with the Arctic, the more distant the Arctic will be, irrespective of topographical geography.

Global Britain in a Global Arctic

Today, two new 'global' concepts are coming to the fore of international politics in ways that are relevant to this book: 'Global Arctic' and 'Global Britain'. The former term relates to the emergence of the Arctic as a region of 'global interest' since the beginning of the twenty-first century.

As I noted in Chap. 2, the term 'Global Arctic' attempts to capture the way in which modern phenomena such as globalisation and climate change brought the Arctic into global currents of science, commerce, and security. The term, which appears to have become the guiding principle behind the annual Arctic Circle Assembly in Reykjavik, embraces interest in the region from everywhere beyond the Circumpolar Arctic, irrespective of topographical geography and history, connecting the Arctic to all corners of the world in the process. Consequently, as the geopolitics scholar Klaus Dodds has suggested, the 'Global Arctic'

> sounds reasonable and timely. But like ice, it can quickly undergo state-change and be reconfigured, reimagined and resorted in ways that benefit some people, places, practices, interests and ideas more than others. (Dodds 2016)

Contrary to the limitless imaginary it attempts to conjure up, the 'Global Arctic' still has a geography—in fact it is likely that it has several geographies depending on whose knowledges and experiences are taken into account, and that in turn means that, as with other 'global phenomena', we should not assume that geographies involving competing sites, knowledges, actors, and materials are no longer relevant.

The more recent emergence of the term 'Global Britain' further threatens to complicate our understanding of how Britain might seek to define and represent its interests and role in the Arctic in the future. The term has gained traction within the British Government over the past year or so as ministers have sought to reframe Britain's expected departure from the European Union (EU) in 2019 as an opportunity to go out and forge new links between Britain and the rest of the world. At Lancaster House on 17 January 2017, in her first major foreign policy speech as Prime Minister, Theresa May mentioned 'Global Britain' more than ten times as she declared:

> The great prize for this country—the opportunity ahead—is to use this moment to build a truly Global Britain. A country that reaches out to old friends and new allies alike. A great, global, trading nation. (May 2017)

Like 'Global Arctic', 'Global Britain' also seems to imply that anything is possible, regardless of history or geography, while avoiding any sort of specificity about how new relationships with the rest of the world will be configured.

If these terms continue to gain currency among foreign policy practitioners, they are likely to become increasingly relevant to the way Britain defines and represents its role and interests in the Arctic. That is significant as they have the potential to form a narrative that justifies a scaling up of British Arctic policy in the coming years. As I argued in Chap. 3, the concept of a Global Arctic poses a substantial challenge to circumpolarisation and the more readily it is deployed, the more likely it is that non-Arctic countries such as Britain will be able to justify a greater presence in the Arctic across a range of commercial, scientific, military, and other interests, particularly those that touch on global issues such as climate change, international law, trade, and the management of global commons. That in turn would speak to the strengths that Britain has traditionally shown as a producer of global science, global financial flows, cutting-edge technology, and international law and governance regimes, which might then lead to a more enhanced leadership role in Arctic affairs—something that the British Government called for explicitly in its 2013 Arctic Policy Framework.

The notion of a Global Arctic also opens up opportunities for Britain to embrace new Arctic partners, beyond the A8 and long-standing European Observers. For example, Britain and China have strikingly similar interests when it comes to Arctic shipping, science, geopolitical stability, and commercial development. The countries' bilateral partnership is strengthening and closer cooperation in the Arctic could lead to joint ventures in science and commerce that bring together the best of British knowledge, expertise, and innovation, with Chinese capabilities and finance.

Lastly, where the Global Arctic meets Global Britain, the Polar Regions Department might find that there is an opportunity to raise the Arctic's profile across the rest of the British Government by drawing attention to the various ways that Britain's broader foreign policy interests will be impacted by what happens in the Global Arctic. Global Britain will require a joined-up global foreign policy, and the various government departments with overseas interests and responsibilities might find it surprising how often they encounter issues that are relevant to what is happening in the Arctic, especially where climate change, scientific endeavours, technological innovation, regime formation, and shifting patterns of trade are involved. That in turn could lead to increased resources for Arctic policy initiatives, especially where they feed the broader goals of Global Britain.

However, the productiveness of the two terms 'Global Britain' and 'Global Arctic' will largely depend on what kinds of geographies they configure, especially in terms of the sites, knowledges, actors, and materials that they connect and mobilise, and neither term will be able to escape responsibility for reterritorialising British–Arctic relations in new ways. Nor can either ever refer to anything that is truly global. Rather, both terms encourage a way of thinking about the geographies of British–Arctic relations as configurable in new ways that transcend the dominance of circumpolarity as the defining principle of geopolitical hierarchy in the Arctic. That would stand in stark contrast to the current approach of assuming that the only way Britain can make itself relevant in the Arctic is by showing due deference to, and support for, the established geopolitical order being enforced by the A8.

If the British Government chooses not to engage with these geographical possibilities, then it is likely that both 'Global Britain' and 'Global Arctic' will add very little to the further development of British Arctic policy. Their beguiling nature obscures the fact that without extensive consideration of what they actually refer to, they remain, simply, ambiguous terms that can mean virtually anything to anyone. That, in turn, could prove useful for a Government which, as Chap. 5 argued, has tended to define British interest in the Arctic in restrictive terms, leaving similar scope for manoeuvre and/or indecision as it has done previously with documents such as the Arctic Policy Framework.

Finding a Role in the Arctic

More than 50 years have passed since Dean Acheson, the former US Secretary of State (1949–1953) under President Truman, uttered the seemingly immortal line that 'Great Britain has lost an empire, but not yet found a role' (Gaskarth 2013: 89). Acheson was speaking to the fact that Britain's ability to act as an independent, globally oriented power in international affairs had diminished substantially since the end of the Second World War. His accusation has continued to resonate in the decades since, resurfacing again and again at key moments of British foreign policy, such as during the Falklands War, as well as during its military adventures in Iraq and Afghanistan. Foreign leaders know it to be the button to push whenever they want to resist British 'meddling' overseas. Inder Kumar Gujral, who at the time was India's Prime Minster, is reported to have rebuffed Britain's offer to mediate between India and Pakistan in their conflict over Kashmir in 1997 by saying Britain was 'a third-rate power nursing illusions of grandeur of its colonial past' (ibid.: 89). More recently, the official spokesman for Russia's President Vladimir Putin described

Britain as 'just a small island ... no one pays attention to them' (Kirkup 2013).

Such 'put-downs' are also relevant in the Arctic where Britain, once an Arctic state itself, gave up its remaining territorial interests in 1920 when it signed the Svalbard Treaty (having already completed the handing over of its North American territories to Canada in 1880). In the decades since, the circumpolarisation of Arctic affairs has pushed Britain to its fringes. Should the Arctic emerge as a new centre of global commerce—a trillion dollar ocean even—connecting Asia, North America, and Northern Europe, Britain risks being among those former imperial centres pushed further still into the periphery of Arctic affairs.

Yet precisely what role Britain should play in the Arctic is still up for debate. Throughout this book, I have deliberately avoided advocating for any particular outcome. Rather, the aim has been to show that what the Arctic is to Britain and what Britain is to the Arctic has always been, is, and always will be open to debate, negotiation, and compromise. The possibilities are wide-ranging and the reality is that Britain will likely play several different roles, which may at times appear contradictory even as it seeks to balance its interests in exploiting the Arctic for economic gain, its scientific capabilities, and its desire to protect the Arctic for future generations. For instance, the Environmental Audit Committee in 2013, along with several leading environmental non-governmental organisations (NGOs) supported by millions of British citizens wanted to see the Arctic, or at least part of the Arctic, turned into a nature reserve. The extractive industries, which contribute billions to the national economy, continue to eye up the Arctic's hydrocarbons, and other valuable minerals and resources, and only need extraction to become profitable before they will try again to turn the region into the next resource frontier. Scientists want to continue testing and investigating the Arctic for clues about climate change and other environmental phenomena, treating the region as the object of their experiments, or at the very least a laboratory for other kinds of research. Sections of the defence community retain an interest in the strategic possibilities presented by the Arctic, especially as, while the ice continues to retreat, the region is becoming permissive of new kinds of military activity, and on a greater scale. Lawyers and insurance firms are working together in the City of London to set new standards of best practice for Arctic industries to protect both people and the environment from the potential negative impacts of development.

But none of these interests will matter if the British Government, businesses, scientists, environmental NGOs, military planners, and others are not prepared to invest in building new connections to the Arctic that further enhance Britain's topological proximity to and presence in the region. It is only through the expansion of such topologies that Britain as a whole can become a source of influence in the region, contesting attempts by others to turn the Arctic into something that might be less desirable from a British perspective, whether by creating new commercial, environmental, or strategic threats to the British mainland, or by denying Britain the possibility of gaining benefits that might arise from changing patterns of trade and development in the region.

We cannot know for sure what will happen next in the Arctic, but all the signs point to greater dynamism as the sea ice retreats, the environment becomes more unstable, the region is developed, and the constellation of actors involved in Arctic affairs expands to include many more from beyond the region. Britain cannot rely on the Arctic being the relatively stable space that it once was. Nor can it wait to see what it will become, as that is likely now to take decades, if not centuries to unfold. Rather, it is a time to be proactive, a time to use greater dynamism and uncertainty in Arctic affairs to pursue what some might consider a more progressive geopolitics, utilising British expertise in producing world-class science, setting international standards for industry, developing leading-edge technological solutions to tackle social and environmental challenges, and creating innovative solutions to emerging problems through greater dialogue and engagement with Arctic states and peoples, as well as others from beyond the region. All of these strengths should be brought to the fore of the public imagination, instead of those moments from Britain's imperial past when the Arctic was simply regarded as something to be subjugated to its will.

Through this kind of progressive sense of place (as the late Doreen Massey, an eminent geographer, might have argued), what Britain and the Arctic mean to each other will be allowed to evolve as new topologies emerge, so that what Britain does now and in the future matters far more than what it did in the past (Massey 1994). There is no reason per se why that should involve any violation of the existing sovereign rights of nations or peoples of the Arctic—investing in dialogue, debate, and material connections with those living in the Arctic is not the same as attempting to subjugate them to the British national interest. Moreover, such an approach would stand in stark contrast to the current approach of memorialising Britain's distant Arctic past (and the imperial legacies that underpin it), while being overly deferential to the present geopolitical status quo that has recently been defined primarily through internal dialogue among the A8.

Alone, Britain cannot determine the Arctic's future, but by building up its connections to the region—by bringing the Arctic closer—it can try to influence it, and in that way perhaps the Arctic also becomes the test bed for a more proactive and focused foreign policy that can avoid charges of 'short termism', 'recentism', and 'strategic drift' that have characterised the British foreign policy establishment of late. But to do so, Britain will first have to answer outstanding questions about what it wants from the Arctic, as well as which connections to invest in both materially and imaginatively, and which to let wither away. Second, Britain will have to acknowledge that there will be some connections that it simply cannot avoid, as the increasingly dynamic Arctic will inevitably make its own way into British life in forms that cannot be easily manipulated or controlled, for instance, through its effects on weather patterns, ecosystems, and trade routes, as well as sources of food and energy.

In both cases, it is perhaps time for British-based social scientists to be given a greater role alongside natural scientists in researching and identifying what Britain's priorities should be in the Arctic. After all, with the rapid retreat of Arctic sea ice, we stand on the cusp of one of the most dramatic changes to the Earth that humans have ever witnessed, and the public conversation about what that means to Britain in the twenty-first century is only just beginning.

Notes

1. For instance, in July 2017, another Arctic exhibition was opened at the National Maritime Museum in London under the title 'Death in the Ice: The Shocking Story of Franklin's Final Expedition'.

References

Depledge, Duncan. 2013. Assembling a (British) Arctic. *The Polar Journal* 3: 163–177.

Dodds, Klaus. 2016. What We Mean When We Talk About the Global Arctic. *Arctic Deeply*, February 18. Accessed January 6, 2017. https://www.newsdeeply.com/arctic/community/2016/02/18/what-we-mean-when-we-talk-about-the-global-arctic

Gaskarth, Jamie. 2013. *British Foreign Policy*. Cambridge: Polity Press.
Kirkup, James. 2013. Russia Mocks Britain, the Little Island. *The Telegraph*. Accessed August 8, 2017. http://www.telegraph.co.uk/news/worldnews/europe/russia/10290243/Russia-mocks-Britain-the-little-island.html
Massey, Doreen. 1994. *Space, Place and Gender*. Minneapolis: University of Minnesota Press.
May, Theresa. 2017. Speech by Theresa May at Lancaster House. *The Telegraph*, January 17. Accessed June 22, 2017. http://www.telegraph.co.uk/news/2017/01/17/theresa-mays-brexit-speech-full/

Index

© The Author(s) 2018

D. Depledge, *Britain and the Arctic*,
https://doi.org/10.1007/978-3-319-69293-7